GEOLOGICAL DISATERS IN THE PHILIPPINES

THE JULY 1990 EARTHQUAKE AND THE JUNE 1991
ERUPTION OF MOUNT PINATUBO

Description, effects and lessons learned

Giovanni Rantucci

The book Geological Disasters in the Philippines was presented for the first time to the World Conference on Natural Disasters, held in Yokohama (Japan) May 23 1994, within the contribution of the Italian Ministry of Foreign Affairs to the International Decade for Natural Disasters Reduction (IDNDR).

Cover: The July 1990 ground rupture (top) in Digdig (Nueva Ecija, Luzon). The picture shows the surface faulting across the road to Carranglan.
Mount Pinatubo eruption plume (bottom) from the East-side of Clerk Air Base. Courtesy of David Harlow (USGS).

PREFACE TO THE SECOND EDITION OF "GEOLOGICAL DISASTERS IN THE PHILIPPINES" (The 1990 earthquake and the 1991 eruption of Pinatubo)

Compared with the previous twenty years (1970-1990), seismicity linked to the subduction zones along the Pacific Rim appears to be on the rise during the last two decades. With its 452 volcanoes, the Pacific Ring of Fire hosts more than 75% of the planet's active volcanoes, 90% of the earthquakes, 81% of which are the biggest at world level. It is worth recalling that since 1990, 20 big earthquakes[1] (> 7.6 Magnitude), a few of which triggered a tsunami, occurred along the Pacific Rim and the nearby zones. Re-publishing twenty years later "Geological Disasters in the Philippines", helps to draw attention to the worsening risk affecting over one billion people increasingly concentrating along and in the vicinity of the Pacifĭc Plate Boundary.

Most scientists believe that the apparent growing seismic activity of the Pacific Plate is the result of (i) the advanced monitoring network providing today more adequate information, and (ii) the economic development of more and more densely inhabited coastal zones. Twenty years after the Luzon Earthquake (1990) and the eruption of Pinatubo (1991), the Pacific Rim Countries should devote more attention to the use of their coastal areas, in which huge industrial and tourism investments are made. Population growth is expected to reach 8 billion within 2025. Part of it will concentrate along the Pacific Rim and nearby zones, thus representing a threat to the rapid expansion of the economies of Asia. The Author wishes to draw attention to four of the 20 earthquakes (eight of which in Indonesia), which occurred since the two geological disasters in Luzon in the early 1990s, and to the lessons learned:

1) In Luzon, the widespread liquefaction (Central Valley) associated with the 1990 earthquake and the damage to structures (Baguio, in the Cordillera Central), the 1991 eruption of Pinatubo and the ash spewed in Luzon, conjured a further disaster: lahars. The rainy season(started in June 1991), heavily affected Pampanga, carrying detrital material produced by the quake and transforming the ash blanket into destructive mud-flows;

2) The 1995 earthquake in Kobe, Japan, showed with unprecedented evidence that the steel reinforcement of buildings and the design of reinforced concrete structures in high-seismicity zones are fundamental;

3) The Andamane Sea sub-marine earthquake of 2004, followed by a most powerful tsunami, caused unprecedented devastation and innumerable casualties, and showed the vulnerability of coastal human settlements;

4) The March 11, 2011 earthquake in Japan, off the Pacific coastal area of Tohoku, the associated tsunami, the thousands of casualties and the damage which affected Fukushima's Nuclear Power Plants are signs of the inadequacy of a development which does not care enough for the safety of people and activities.

---

[1]List of the 20 biggest earthquakes which occurred mostly along the Pacific Ring of Fire, and a minor quantity in the nearby areas: (1) The Philippines, Luzon, July 16, 1990, M 7.7; (2) Indonesia, Flores Island, December 12, 1992, M 7.7; (3) Japan, Hokkaido, June 12, 1993, M 7.7; (4) Indonesia, February 17, 1996, M 8.2; (5) Peru, November 12, 1996, M 7.7; (6) Indonesia, June 4, 2000, M 7.9; (7) Papua New Guinea, November 16, 2000, M 8.0; (8) Peru, Southern Region, June 23, 2001, M 8.4; (9) Japan, September 29, 2003, M 8.3; (10) Indonesia, Andamane Sea and Nicobar Islands, December 26, 2004, M 9.3, plus tsunami; (11) Indonesia, Sumatra, March 28, 2005, M 8.6; (12) Chile, June 13, 2005, M 7.8; (13) Indonesia, Java, July 17, 2006, M 7.7; (14) Solomon Islands, April 1, 2007, M 8.1; (15) Indonesia, December 9, 2007, M 8.5; (16) Chile, November 14, 2007, M 7.7; (17) Samoa, September 29, 2009, M 8.1; (18) Chile, February 27, 2010, M 8.8; (19) Japan, March 11, 2011, M 9.0, plus Tsunami; (20) Indonesia, Banda Aceh, April 11, 2012, M 8.7 (Tsunami alert was raised across the Indian Ocean Region). The epicenter is located some 500 km south-west of Banda Aceh in the Indian Ocean. Source of data concerning the first 19 events is the List of Deadly Earthquakes since 1900 – Wikipedia.

FOREWORD

Geologic phenomena including earthquakes, volcanic eruptions, land-slides and erosional processes are an essential aspect of the evolution of the environment in which we live. The exponential growth in the earth's population that occurred in this century and the associated development have progressively resulted in the overcrowding of several vulnerable areas of the planet. Thus, the occurrence of extreme natural phenomena increasingly entails disasters with loss of numerous lives, damage to infrastructure and property, and widespread human suffering. In this sense disasters, which more and more often result from the combined effect of human development and natural forces, substantially contribute to the global environmental crisis.

The book «Geological Disasters in the Philippines» is centered on these views and is designed to improve the understanding of the recent extreme geological events which affected the Archipelago in the early 1990s. A considerable part of the work is devoted to the description of disaster impacts on physical and human environments as well as on agriculture and the economy of the Philippines.

A great deal of knowledge about geological disasters in the Italian Peninsula (and more generally in the Mediterranean Region) was gathered as a result of the abundant literature from the Roman Times onward. It is worthwhile to mention the classic work of the latin writer Lucius Annaeus Seneca (5 B.C.-A.D. 65) «De terrae motu» (On the Motion of the Earth) inspired by the destructive earthquake which struck the town of Pompeii in A.D. 62 or 63. Pompeii and Herculaneum were buried by volcanic ashes in A.D. 79, that is about 17 years later, by the explosion of Mount Somma and Vesuvius. This disaster was described by the historian Pliny the Younger (A.D. 62 - 114) who wrote the first scientific report of a volcanic eruption.

The Italian experience with disasters was greatly enlarged in this century by the recurrence in the country of a number of calamities. Beyond the high-quality data which were gathered, these disasters triggered advanced research in various fields. The descriptions by historians during several centuries combined with the information from recent research constitute the best documented data bank on disasters during the last two millennia.

In view of the recurrent threat posed by extreme geological events, Italy has recently made considerable efforts in disaster preparedness, prevention and mitigation by adequately organizing the social response.

Italian Institutions are actively participating in the IDNDR initiatives for a worldwide reduction of disaster impacts, and contribute to alleviating the effects of calamities by sharing knowledge, available data and technical know-how.

Investments by industrial countries are essential for the protection of efforts by developing countries towards economic growth and sustainable development. The Italian Development Cooperation has actively supported Developing World in the last 15 years, contributing also to disaster prevention and mitigation projects such as SEISMED, which was designed to reduce seismic risk in the Mediterranean Region.

The Author of the book, Dr. G. Rantucci, has worked for over ten years as a geologist in Asia, spending part of this period in the Philippines and a major part as Associate Professor at the Asian Institute of Technology in Bangkok. At present he is an Expert in the Technical Unit of the Italian Ministry of Foreign Affairs, Directorate for Development Cooperation. This book reflects the variety of his experience and the versatility of his expertise.

Antonio Catalano di Melilli
Deputy-Director General
Directorate General for Development
Cooperation
Italian Ministry of Foreign Affairs

PREFACE

Dr. Rantucci was one among many foreign scientists who were in the Philippines during the world class disasters that hit the Country in 1990 and 1991 - the July 16, 1990 Luzon earthquake and the 1991 eruption of the Pinatubo Volcano. He is also one of the few foreign scientists who were inspired by these two disasters to write and publish papers, but he is the only one who has produced a full monograph so far.

In the Philippines, we produced several compilations containing papers written on these two disasters by both local and foreign scientists including Dr. Rantucci. These compilations are:

- The July 16, 1990 Luzon Earthquake: a Technical Monograph;
- Proceedings of GEOCON '90, Quezon City, Philippines 5-7 December 1990;
- Proceedings of the International Scientific Conference on Mt. Pinatubo, Manila, Philippines 27-30 May 1992;
- Proceedings of GEOCON '91, Quezon City, Philippines, 4-6 December 1991.

However, only limited copies of these publications were printed and circulated locally. A technical monograph on the Pinatubo Volcano 1991-1992 eruptions and their aftermath is also in the making - a joint effort of the United States Geological Survey (USGS) and the Philippine Institute of Volcanology and Seismology (PHIVOLCS). This will have a wide international circulation but covers only the Pinatubo Volcano eruption. As a document integrating both disasters for international circulation, Dr. Rantucci's book is therefore a first.

Dr. Rantucci's book will certainly make our country famous internationally. We hope though that the image that will stick in the readers' mind will not be of a country prone to, and hard hit by disasters, where only the brave visitors and investors dare to tread. Rather, the image that should last in readers' memories is of a disaster-prone country whose leaders and citizens have learned their lessons well and taken steps toward effective disaster prevention and mitigation.

Raymundo S. Punongbayan
*Director of the Philippine Institute
of Volcanology and Seismology*

ACKNOWLEDGEMENTS

The publication of this book was sponsored by the «Dipartimento per l'Informazione e l'Editoria», a Government Institution attached to the Office of the Italian Prime Minister.

A particular thank is deserved to Senator Edmundo Mir, Undersecretary at the Department of Public Works and Highways of the Philippines for inviting the Author to Dagupan City, a town severely affected by the July 1990 earthquake, and on that occasion triggering his interest in the extreme geological phenomena hitting the Archipelago.

The Author feels deeply grateful to Dr. Raymundo Punongbayan, Director of the Philippine Institute of Volcanology and Seismology for his support to the work in terms of suggestions, data, published papers, information and pictures. Without the cooperation with PHIVOLCS and the excellent work of its scientists and researchers the publication of this book would not have been possible.

The Department of Public Works and Highways (DPWH) and the Bureau of Soils and Water Management (BSWM) of the Philippines were important sources of information and data.

Special thanks to Prof. Domenico Giardini and particularly to Dr. Laura Beranzoli of the Istituto Nazionale di Geofisica (ING), who kindly provided help, data and suggestions.

The author is very grateful to Dr. Robert Brinkman for his critical comments, the revision of the text and the invaluable enthusiasm which accompanied his work during several months.

Thanks are expressed to Organizations and Authors which kindly granted permission to use or reprint their material.

Comments, remarks and suggestions were provided by Dr. Franco Maranzana, Dr. Domenico Bruzzone and Mr. Manuel Goseco.

A particular gratitude goes to my family, friends and colleagues for the support during more than two years of work.

ABOUT THE AUTHOR

Giovanni Rantucci received his Degree in Geology from the University of Rome (Italy) in 1964. During an assignment as a consultant in Central Luzon (Philippines), he witnessed the sequence of geological events described in the book: the July 1990 Luzon earthquake, the June 1991 eruption of Mount Pinatubo and the related primary and secondary effects.

During the period 1983-89 the Author served as Associate Professor at the Asian Institute of Technology in Bangkok (Thailand), Division of Geotechnical and Transportation Engineering, in a multilateral project financed by the Italian Cooperation (Ministry of Foreign Affairs) through the UNDP. Previously (1968-1983) he worked as a geotechnical engineer for a private Italian company, being involved in numerous projects in the Middle East, Far East, Africa and South America. At present he is a member of the technical staff of the Italian Ministry of Foreign Affairs (Rome) in the Directorate General for Development Cooperation.

CONTENTS

Chapter 1

NATURAL HAZARDS AND DEVELOPMENT

1.1 Introduction

The idea of writing this book originated during the occurrence of a sequence of natural disasters on the island of Luzon (Philippines). These included the July 1990 earthquake, the June 1991 Mount Pinatubo eruption and the related primary and secondary effects. The two disasters, their consequences on the economy of the Philippines and the lessons learned are described in detail.

An important aspect of the book is the connection between disasters, human development and the environment. Human activities in general have a major change potential on the troposphere and their influence is unique since it consists of a multitude of local impacts simultaneously affecting a wide range of natural realms. Human use and manipulation of environment and resources can significantly alter natural processes and aggravate the impact of extreme natural phenomena.

An example of how anthropic activities can cause major environmental change has been provided by Crosby (1992) in his research on the biological and cultural consequences (and vice-versa) of the European colonization of the Americas. In the case of the Philippines the widespread deforestation during the last decades in Luzon heavily contributed to some of the devastating effects of the 1990-91 disasters.

Description and analysis of the natural phenomena that occurred in Luzon, their strict sequence and effects on human beings, property and the country's economy should help researchers, geoscientists, decision makers, administrators and planners in various ways. Through the description of the 1990-91 geological events and the impacts associated with them, this book should contribute to enhance the knowledge on natural disasters and improve society's response to them. It seeks also to draw special attention to the need for the introduction of natural hazard assessment and mitigation measures into the development planning process as a major step in risk reduction.

The general framework of disasters, as well as interactions with the present global crisis are described in Chapter 1. Past disasters in the Philippines and recent events in Luzon are summarized in Chapter 2. Tectonics and Geology of Luzon, with emphasis on aspects directly related to the recent disasters, are described in Chapter 3. Chapters 4, 5, 6 and 7 deal with the 1990 earthquake, liquefaction in the Central Plain, landslides in Central Cordillera and Caraballo Mountains, and the aftershock swarm, in that sequence. Chapter 8 is devoted to the Pinatubo eruption and its consequences in Central Luzon. Chapter 9 gives an overview of damage caused by the events of 1990-1991 and their effects on the economy of the Philippines. Chapter 10 describes the lessons learned from Luzon's disasters, focusing on general as well as on specific aspects.

1.2 Extreme phenomena and natural disasters

Extreme natural phenomena may occur anywhere in the world and every area of the planet has probably been affected by some of them during part of its geological history. When extreme events take place in a populated area they are referred to as hazards and termed disasters if they cause numerous casualties, widespread human suffering and considerable property damage. According to UNESCO a disaster occurs when a natural hazard seriously disrupts the functioning of a community, causing widespread human, material and environmental losses that exceed the community's capability to cope without external relief. As the human population increases disasters became more frequent.

Various natural phenomena that turn into disasters are characterized by the occurrence of isolated peak-events, in comparison to the regular magnitude and frequency of the majority of episodes of the same nature: typical is the case of a strong earthquake with a recurrence period of centuries for instance, compared to the myriad of small and harmless tremors that normally hit the same area. Most common disasters include earthquakes, volcanic eruptions, tsunamis, landslides, windstorms, floods, droughts, wildfires, insect infestation and the gradual but progressive degradation of natural resources. Some of them occur suddenly, others are advanced by early signs, others take place slowly.

The majority of earthquakes, tsunamis, and floods belong to the first group (the July 1990 quake, for example, struck Luzon suddenly). The 1991 eruption of Pinatubo, which belongs to the second group, was heralded a few months earlier by premonitory signs, thus the volcanic district was under observation long before the critical episode. A number of disasters, by contrast, belong to the third group since they display their effects during long timespans. Prolonged drought, or the progressive degradation of the land at a regional level, due to deforestation, soil erosion, and inappropriate land use, are typical examples.

Compared to natural phenomena that can recur after decades ar centuries (for example strong earthquakes), some disasters related to climate are characterized by short return periods. Typhoons hit tropical regions with frequencies estimated in years, increasingly provoking disasters as the affected areas become more densely populated and human activities more widespread.

Disasters are characterized by primary effects and in many cases they also have secondary, medium to long-term consequences. The primary effect of the Mount Pinatubo 1991 eruption in the Philippines consisted of the deposition of a vast ash blanket throughout Central Luzon with severe consequences for human activities and for the natural environment. The secondary medium- to long-term impacts were caused by devastating mudflows during the rainy seasons that followed the eruption. In this case the secondary effects were less destructive than the eruption, but the damage to agricultural and other productive activities induced by the recurring mudflows has continued for years.

Climate, geological phenomena and their combined effects are sources of disasters in numerous areas of the planet. In this context, the Caribbean and the Philippine-Indonesia regions share a most hazardous location.

Figure 1.1 shows a world map of natural hazards (1988), classed as two groups: the first includes earthquakes, tsunamis and volcanic eruptions (related to lithosphere dynamics), the second consists of tropical storms and cyclones, winter gales and tornadoes (related to climate). The map appears to over-emphasize the zone affected by tropical cyclones, since most of the area in violet is ocean where the population is minimal; by contrast, it does not portray flood and drought hazards, which affect many more people than cyclones and earthquakes do.

Natural hazards are not likely to change in the near future, thus their destructive potential will definitely increase. United Nations predictions indicate that by the year 2000 there will be worldwide 31 megacities of over 8 million people and 67 smaller cities with 2.5 to 8 million people. The on-going massive urbanization and industrialization processes are likely to increase the vulnerability of populations and human activities in the areas at risk.

1.3 Disasters and developing countries

The cost of the damage caused by natural disasters, which was estimated at about US$ 400 million during the 1950-1959 period, increased ninety times to 36 billion in the decade 1980-1989 (World Bank, 1990).

Further Natural Hazards, Other

- △ △ △ Limit of iceberg drift
- Temporary pack ice
- Permanent pack ice
- Sea fog frequency above 30% (July)
- Isoline of thunderstorm days per year
- □ Bombay — more than 1 million inhabitants
- ○ Chembote — 100 000 to 1 million inhabitants
- ○ Townsville — less than 100 000 inhabitants
- ● Bonn — capital city
- □ Sydney — MR office abroad
- State borders (These should not be regarded as official.)
- Rivers

Windstorms

1. Tropical storms and cyclones (Beaufort 8 and above)
- 0.1 bis 0.9 per year
- 1.0 bis 2.9 per year
- 3.0 and more per year
- Isoline of maximum frequency (Arabian Sea: monsoon gales)
- Average tracks

2. Winter gales
- Per cent frequency of Beaufort 7 and above: North Atlantic and North Pacific: December; Southern hemisphere and Arabian Sea: June
- Isoline of per cent gale frequency

3. Tornadoes
- Number of symbols per major area: average frequency per year
- USA: isoline of tornado frequency, in centuries (eg 50 = "return period" of 5,000 years per location)

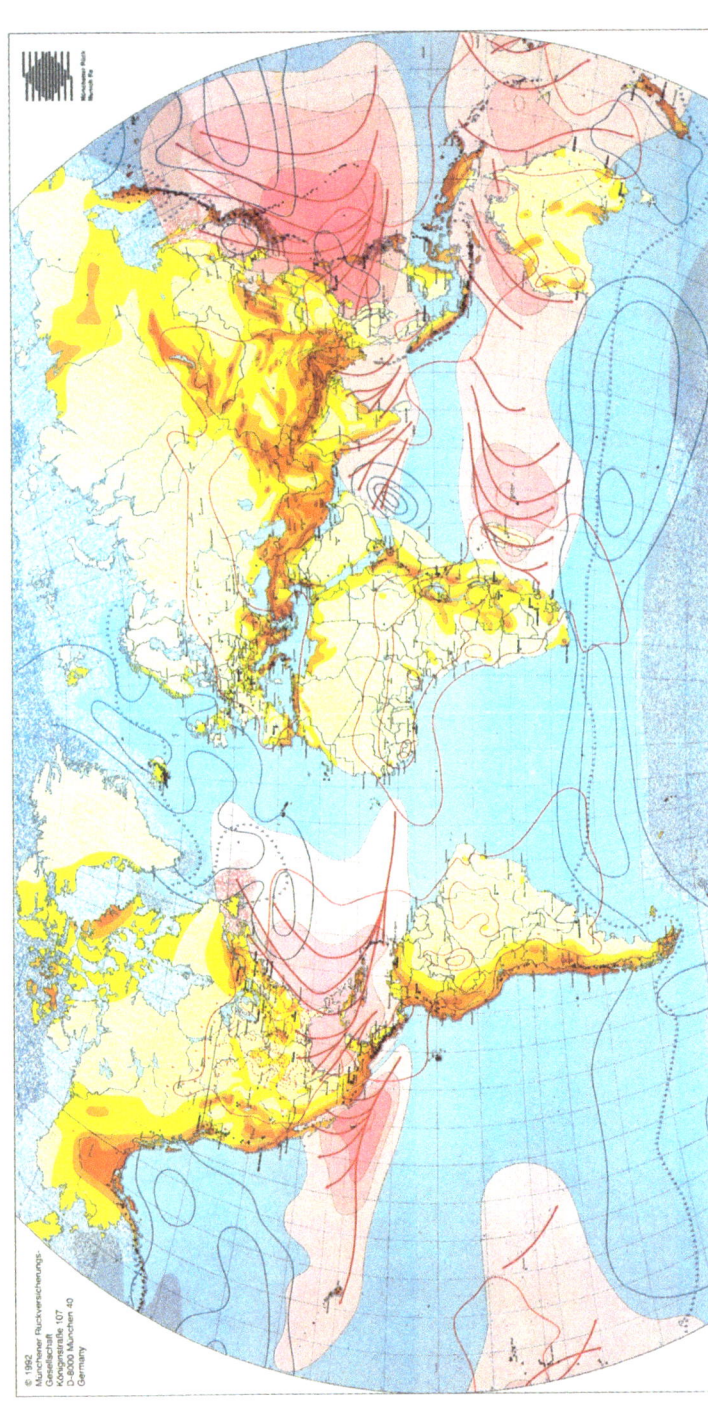

Earthquakes, Tsunamis and Volcanoes

Probable maximum intensity (Modified Mercalli Scale MM) with an exceedance probability of 20% in 50 years equivalent to one occurrence probably 250 years ("return period") on average, for medium subsoil conditions:

- Zone 0: MM V and below
- Zone 1: MM VI
- Zone 2: MM VII
- Zone 3: MM VIII
- Zone 4: MM IX and above
- Coasts exposed to tsunamis (seismic sea waves)
- ▲ Active volcanoes
- ▲ High-risk volcanoes

Earthquake Intensity Scales

MM 1956	Descriptive term	Acceleration %g	MSK 1964	RF 1883	JMA 1951
I	Imperceptible	<0.1	I	I	0
II	Very slight	0.1–0.2	II	II	I
III	Slight	0.2–0.5	III	III	II
IV	Moderate	0.5–1	IV	IV–V	III
V	Rather strong	1–2	V	VI	IV
VI	Strong	2–5	VI	VII	V
VII	Very strong	5–10	VII	VIII	
VIII	Destructive	10–20	VIII	IX	
IX	Devastating	20–50	IX	X	VI
X	Annihilating	50–100 (≈1g)	X	XI	X
XI	Disaster	1–2g	XI		
XII	Major disaster	>2g	XII	XII	VII

MM 1956 Modified Mercalli; MSK 1964 Medvedev-Sponheuer-Karnik; JMA 1951 Japan Meteorological Agency; RF 1883 Rossi-Forel

Earthquake: Scales and Effects

Earthquake effects and source model: Fault, Epicentre, Hypocentre, Displacement, Fault plane, Landslide — Shaking and fire losses — Liquefaction (sand crater) — Tsunami (seismic sea waves)

Earthquake and Tsunami Magnitude Scales

Earthquake magnitude (according to Richter, 1956)

$$\log E = 11.8 + 1.5\,M$$

E = energy released (in erg): to be multiplied by 32 for each full M grade
M = Richter magnitude (up to M = 9)

In addition to M, effects observed on the surface (= intensities) depend mainly on the depth and the distance from the focus, the duration of the earthquake and the prevailing subsoil conditions.

Tsunami magnitude (according to Iida, 1970)

Grade	Descriptive term	Water level in m
0	Slight	0–1
1	Moderate	1–3
2	Strong	3–6
3	Very strong	7–20
	Disaster	20–

Windstorms: Scales and Effects

Beaufort Scale

Bft	Descriptive term	m/s	km/h	mph	knots
0	Calm	0–0.2	0–1	0–1	0–1
1	Light air	0.3–1.5	1–5	1–3	1–3
2	Light breeze	1.6–3.3	6–11	4–7	4–6
3	Gentle breeze	3.4–5.4	12–19	8–12	7–10
4	Moderate breeze	5.5–7.9	20–28	13–18	11–15
5	Fresh breeze	8.0–10.7	29–38	19–24	16–21
6	Strong breeze	10.8–13.8	39–49	25–31	22–27
7	Near gale	13.9–17.1	50–61	32–38	28–33
8	Gale	17.2–20.7	62–74	39–46	34–40
9	Strong gale	20.8–24.4	75–88	47–54	41–47
10	Storm	24.5–28.4	89–102	55–63	48–55
11	Violent storm	28.5–32.6	103–117	64–72	56–63
12	Hurricane	32.7–	118–	73–	64–

Saffir Simpson Scale (up to hurricane force)

SS	Descriptive term	m/s	km/h	mph	knots	Wind pressure kgm²
1	Weak	32.7–42.6	118–153	73–95	64–82	0
2	Moderate	42.7–49.5	154–177	96–110	83–96	0–0.1
3	Strong	49.6–58.5	178–209	111–130	97–113	0.2–0.6
4	Very strong	58.6–69.4	210–249	131–155	114–134	0.7–1.8
5	Devastating	69.5–	250–	156–	135–	1.9–3.9

Fujita Scale (up to tornado force)

FS	Descriptive term	m/s	km/h	mph	knots	Wind pressure kgm²
0	Weak	17.2–32.6	62–117	39–72	34–63	4.0–7.2
1	Moderate	32.7–50.1	118–180	73–112	64–97	7.3–11.9
2	Strong	50.2–70.2	181–253	113–157	98–136	12.0–18.3
3	Devastating	70.3–92.1	254–332	158–206	137–179	18.4–26.8
4	Annihilating	92.2–116.2	333–418	207–260	180–226	26.9–37.3
5	Disaster	116.3–	419–	281–	227–	37.4–50.5

Fig. 1.1 – *World map of natural hazards (Munich Reinsurance Company, 1988).*

Developed countries have already proved able to cope with many types of vulnerability of the physical and human environments to disasters. In a number of cases government authorities have successfully and cost-effectively dealt with the problems created by extreme natural phenomena by a combination of prevention, reconstruction and rehabilitation programs. The number of casualties was limited, for instance, during the January 1994 earthquake in Northridge, near Los Angeles, where experts had learned much about construction techniques and strengthening of old structures. The earthquake which struck Friuli (northeastern Italy) in 1976 was followed by complete reconstruction after the adoption of adequate techniques and standards.

Developing countries do not have in general sufficient financial resources so that the disaster emergency period can be followed by post-disaster havoc. In some cases the devastated area and its feeble economy do not succeed in recovering and thus revert to higher poverty levels. Whenever the affected country has limited financial resources, the public expenditure is usually redirected to rehabilitation and reconstruction projects, leaving most of the pre-disaster problems unattended. A number of poor countries urgently needs foreign help to reduce losses from natural disasters and strengthen the difficult path towards a sustainable development.

The economies of developing countries are the most heavily affected by the immediate impact on physical assets and employment and by the consequent slowing down of their development. The physical effect of hazardous events is often exacerbated by fragile environmental conditions. Poverty, high rates of population growth, environmental destruction through deforestation and inappropriate land use, inadequate infrastructure and policies, and lack of investments are often associated with this scenario.

According to Anderson (World Bank, 1990) development in some cases increases disaster proneness, and poor countries are comparatively more vulnerable since the environment has often already undergone significant degradation and a sizable depletion of resources has occurred. The effect of local development projects that attract more and more people to areas prone to seismic, volcanic or flooding hazards, can aggravate the consequences of natural disasters. Thus, it can be concluded that vulnerability to disasters needs to be considered as an essential aspect in development programs. Disaster costs can be reduced by proper and timely investments, centered on prevention and mitigation. Despite the huge destruction caused by recent disasters in the Philippines, a significant reduction of damage and casualties was achieved by the work of government organizations (such as NDCC, PHIVOLCS and others) with a background in early warning, organization of disaster emergency, preparedness and mitigation.

1.4 Human development and global hazard

Life and the environment on the planet have been conditioned by a number of phenomena that have recurred in the past. Impacts from the outer space (over 120 astroblemes have been discovered on Earth), critical plate tectonics periods and sudden variations in climate and the environment are natural phenomena with an immense destructive potential.

Whereas these events occurred over a period of several hundred million years, some alarming phenomena associated with the recent human development have become evident today, such as the loss of biodiversity or pollution, and some others will develop in the near future. These include the green house warming and the ozone layer depletion. In the course of the next century heat trapping due to the increasing presence of green house gases can reach much higher levels compared to the preindustrial period. The resulting global warming can significantly influence climate and have severe effects on human development.

Since the 1985 discovery of the ozone reduction over Antarctica, ozone depletion has gone on faster than expected and harmful effects on human beings and life may begin to occur during the next decades (World Development Report, 1992).

Despite the great efforts of scientists to model the earth's climate, the variety of scenarios and uncertainties about feedback effects still preclude prediction of the conditions that might be expected during the next 20-30 years. Many questions are still unanswered, while scientists strongly disagree with each other over some major issues. A number of facts, however, are clear:

– the natural resource base of the planet is finite; however, the consumption/production patterns shaped by the current global economic relationships lead to increasing demand and consumption rates even though these are likely to prove unsustainable.

– world population is now about 5.4 billion. According to the World Development Report (1992), between 1990 and 2030 world population is expected to grow by 3.7 billion, while the need for food will double and industrial production and energy consumption will be three times greater. This implies heavier and more widespread impacts on nature, the environment and resources.

– poverty, political uncertainty and local wars will most probably continue during the next century and negatively affect developed and developing countries.

Thus, many of the present problems related to resource depletion and degradation, poverty and insecurity will increase in magnitude. Under these conditions, the occurrence of natural disasters is likely to have a far more dramatic impact than at present on people, the environment and life in general.

1.5 Disaster distribution and effects during the last decades

The trend towards a sustainable development strategy certainly calls for an integrated approach, since not only do the effects of disasters need to be reduced but also harmful global impacts, such as the human-induced strengthening of the greenhouse effect and the depletion of the ozone layer, the responsibility for which rests mainly with the developed countries. Figure 1.2 shows the evolution of disasters during recent decades.

According to Boutros Boutros-Ghali, United Nations Secretary-General, in constant (1990) dollars economic losses due to disasters have tripled in the last thirty years: $40 billion in the 60s, $70 billion in the 70s, $120 billion in the 80s. Without disaster reduction measures the cost will rise to at least $280 billion in the 1990s (Stop Disasters, N. 17, 1994).

An assessment of natural disasters on earth, over a 25 year period (1966-1990) is presented in the May-June 1993 issue of Stop Disasters. Figure 1.3 shows (top left) the number of deaths due to high winds, earthquakes, floods and volcanic eruptions and the number of people affected (top right). Floodings account for three quarters of the number of people generally affected by disasters; this percentage reflects their great destructive potential. The bottom part of the figure illustrates the Spatial Risk Ratio (SRR) of Countries by type of disaster: New Zealand, Bangladesh and Philippines are the three countries most affected by floods, while Philippines and Bangladesh share the most hazardous locations as far as high winds are concerned. The SRR might be more meaningful if applied within countries rather than between countries: floods, cyclones and other types of disasters generally affect specific areas, not whole countries or populations.

1.6 International decade for the reduction of natural disasters

The correlation between poverty, population growth, degradation of the environment, depletion of natural resources and deterioration in the quality of life is a major issue for present and future generations.

Emerging environmental ethics, ecological associations, Green parties, ecologically oriented NGOs and individuals play a prime role in the transition from indiscriminate growth to equilibrium and the reduction of consequences of natural hazards.

At its 44th session (December 1989), the General Assembly of the United Nations proclaimed the International Decade for Natural Disasters Reduction (IDNDR, 1990-2000), and a variety of steps were taken in various fields, at national and international levels.

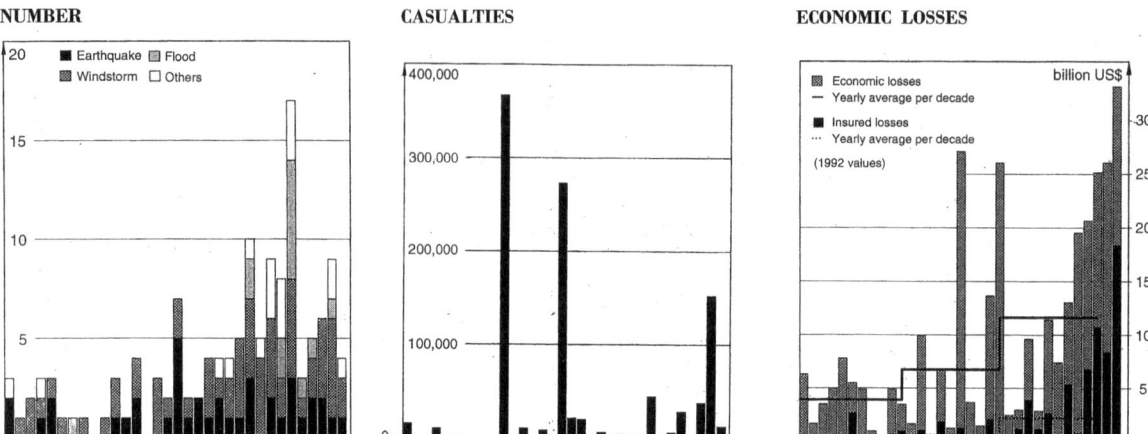

Fig. 1.2 – *Disasters during the last decades (number, casualties and economic losses), Geoscience Research Group, Munich Reinsurance Company (Stop Disasters, N. 17, 1994).*

The initiative «has the objective of substantially cutting the impact of natural hazards in human casualties, property damage and social and economic disruption» (UN/IDNDR, Report 1990-91, United Nations Declaration, 1988). Thus, the general objective of the Decade is the reduction of human and economic losses resulting from the occurrence of natural hazards. By the year 2000 all countries should have achieved:

– comprehensive national assessments of risks from natural hazards, with these assessments taken into account in development plans;

– mitigation plans at national and/or local levels, involving long-term prevention and preparedness and community awareness;

– ready access to global, regional, national and local warning systems and broad dissemination of warnings.

An important aspect links the 1992 Earth Summit in Rio and the 1994 Yokohama Conference on Natural Disoster Redaction: the need to halt and reverse environmental degradation and promote environmentally sustainable development. Reduction of disaster impacts can be considered a first step in meeting some of the major requirements of the Agenda 21 action plan and an investment for future generations.

1.7 Conclusion remarks

Natural phenomena as such do not necessarily carry negative implications, and their change potential is the force through which the planet has evolved through geological times. As a result a planet with life and a balanced environment has been delivered to us. Human activities have recently started altering the environment in more major ways than ever before.

Since then extreme phenomena, which are inherent aspects of natural evolution and which also repeatedly occurred in the past, have begun to turn into disasters. Major natural events are beyond our control, therefore human development has to be modified and addressed in such a way as to minimize the impacts of disasters.

The first lesson to be learned is that human responses to natural hazards need to be integrated into the development planning of present and future generations. The second major aspect is that vulnerability of human settlements, structures and working activities can be reduced by a proper management of natural hazards.

Disaster Ranking Over 25 Years

Total deaths and affected by type of disasters
1966 - 1990

Data from CRED Disaster Events Database (EM-DAT)

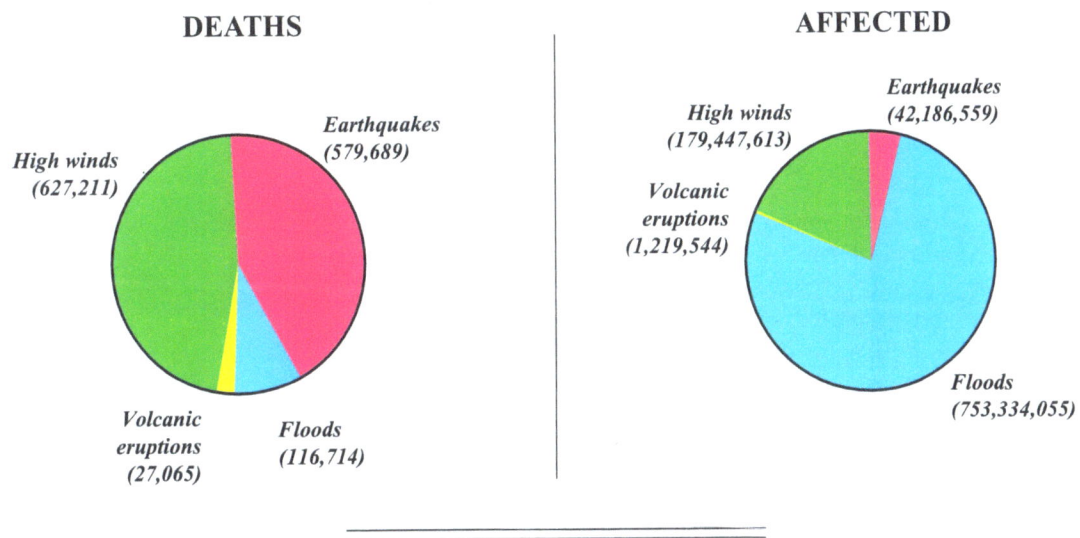

DEATHS

Earthquakes
(579,689)

High winds
(627,211)

Volcanic
eruptions
(27,065)

Floods
(116,714)

AFFECTED

Earthquakes
(42,186,559)

High winds
(179,447,613)

Volcanic
eruptions
(1,219,544)

Floods
(753,334,055)

Spatial Risk Ratio of Countries by Type of Disasters

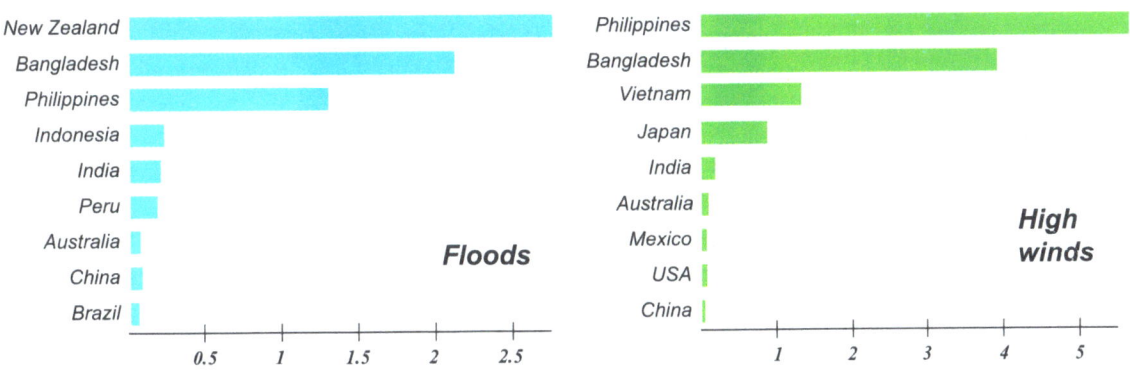

Fig. 1.3 — *Disasters ranking over the period 1966 - 1990. Spatial Risk Index has been calculated by dividing the number of disasters by land and water area of the affected country since the larger the land surface, the higher the overall probability of exposure to geological and meteorological events (Stop Disasters, N. 13, 1993).*

CHAPTER 2

OVERVIEW OF PAST AND RECENT DISASTERS IN THE PHILIPPINES

2.1 Introduction

The Philippine Archipelago is a cluster of 7,107 islands with an area of 0.3 million square kilometers and a coastline over 17,000 km long (Fig. 2.1). Located along the western rim the Pacific Ocean, the country has a population of about 60 million with a annual growth rate of 2.3%. The two major islands of the archipelago, Luzon and Mindanao, cover 65% of the total land area, while Luzon alone has a surface of about 120,000 square kilometers, most of which is mountainous.

The Philippine Archipelago has a humid tropical climate with an average temperature range of 24-35 degrees C and midday summer peaks over 38 degrees C. Humidity is usually high and rainfall ranges from 2,000 to 4,000 mm/year depending on the region. On western Luzon 80% of the total annual rainfall occurs from July through September, while the same percentage affects eastern Luzon from February through May. A three-storied type rainforest originally covered the lowlands of the archipelago, the dominant trees mainly belonging to the Dipterocarp family. Deforestation be-

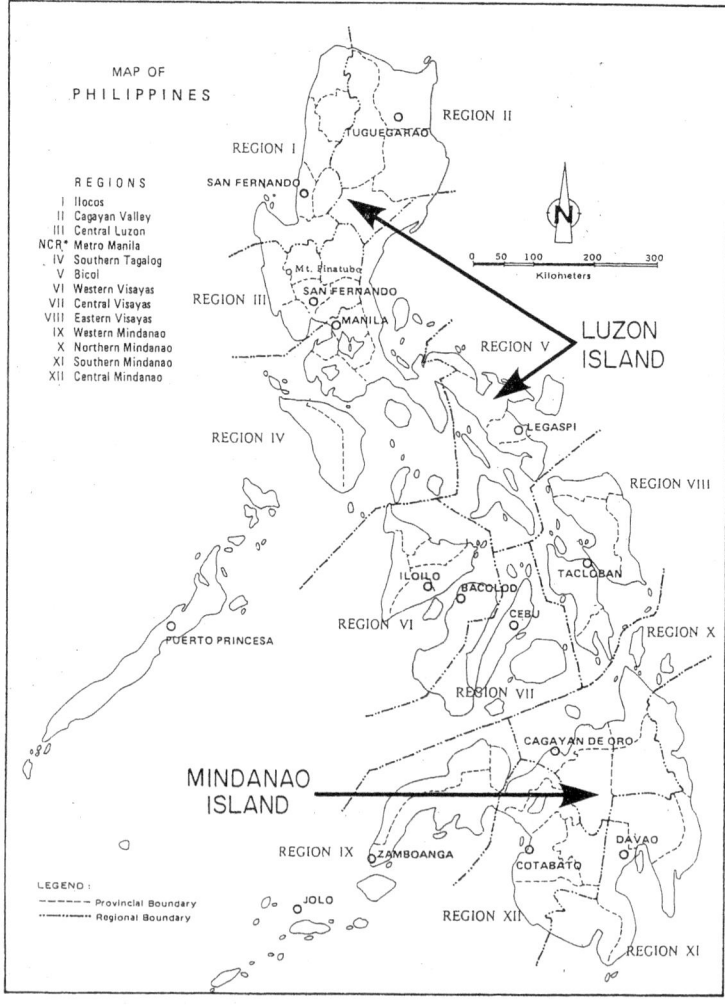

Fig. 2.1 – *The Philippine Archipelago (ADB, 1991b).*

came extensive after the Second World War, and little of the lowland is still forested at present. By the end of this century only the original vegetation of higher-altitude zones may have survived.

The tectonic setting and climate make the Philippines highly vulnerable to different types of hazards. The convergence of the Eurasian and Pacific Plates and the monsoon climate are mainly responsible for the frequent disasters which hit the archipelago.

Numerous calamities of varying nature were documented by Spanish chroniclers since the middle of the 17th century. Earthquakes, volcanic eruptions, landslides, mudflows, tsunamis, typhoons and floods still plague the country, as they did in the past.

Although natural disasters occur in many parts of the world, the Philippine Archipelago is a major target for quite a range of them (Chapter 1, Fig. 1.1).

2.2 The framework of geological disasters in the Philippines

2.2.1 General

A fundamental recognition in natural sciences is that the earth is a dynamically evolving body. The outer layer of the planet, the lithosphere (about 100 km thick), has been continuously subjected to structural and morphological changes throughout geological times. The physical activity of the earth's crust is partly the result of the energy supplied as heat from the earth's interior, partly as energy provided by the sun in the form of radiation. The combined effects of the related forces resulted in a sequence of evolutionary processes which have shaped the lithosphere for hundreds of millions of years.

The phenomena resulting from the aforementioned sources of energy are known as the geologic cycle. The concept, which was introduced by James Hutton in 1785, includes two major components, or subcycles (Fig. 2.2):

— the hydrologic cycle, which is the cyclical path of water from the oceans into the atmosphere and then back to the oceans through precipitation and flowing along rivers and streams. This complex path is responsible for the dismantling of high ground and the downward transportation of sediments towards flat-land areas and marine floors. In connection with the incoming solar energy, the cycle drives the chemical reactions and physical processes which are essential to the alteration and degradation of superficial rocks and to the development and evolution of biological processes.

— the tectonic cycle, which consists of processes related to the heat stored in the earth's interior; these include the continuous deformation of the crust, the motion of the lithospheric plates, the upheaval of mountain ranges, the rise of basaltic magma along oceanic fractures, the intrusion of molten rock into volcanic chambers and the consumption of the oceanic crust through subduction. The motion of continental and oceanic slabs, better known as plate tectonics is responsible for

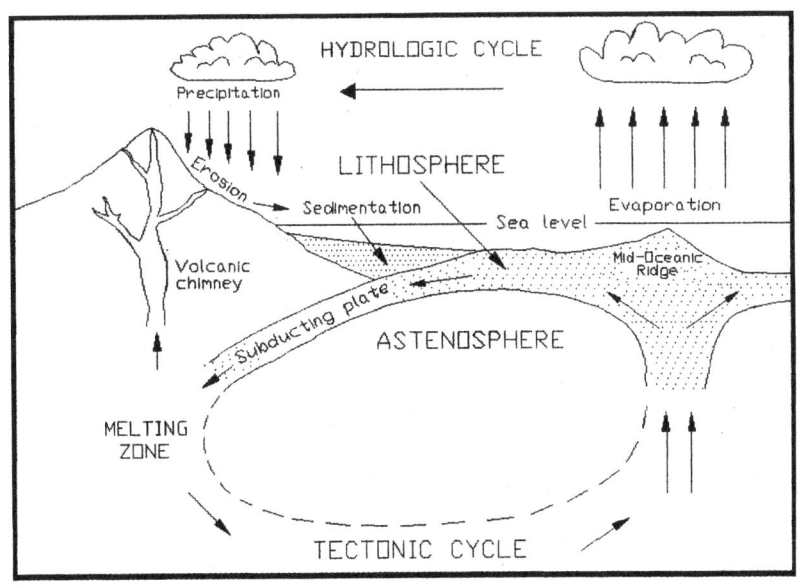

Fig. 2.2 – *The geologic cycle and the hydrologic and tectonic subcycles.*

the collision and breaking apart or separation of plates. Numerous earthquakes and volcanic eruptions associated with the motion of plates occur along the boundary zones between them.

The geologic cycle is characterized, at the local level, by prolonged periods of calm and routine processes, suddenly interrupted by episodes of highly dynamic activity. The troposphere (the interface where life and human development are concentrated) represents the physical space where tectonic and hydrologic sub-cycles widely exert their hazardous influence.

The disasters which occurred in the Philippines during 1990 through 1991 are due to the descent of eastern and western sea floors beneath the Archipelago and the ensuing motion of crustal blocks along the numerous faults in the area. High seismicity and volcanism are associated with these dynamic processes: the numerous and strong earthquakes which have been hitting the Philippines, since remote geological times, mainly originate near and along plate boundaries, subduction zones and the Philippine Fault, while eruptions occur along active volcanic lineaments. The seismicity of the Archipelago is also responsible for a number of tsunamis along the coastal areas, landslides in mountainous provinces and liquefaction in the plains. The latter phenomenon, in turn, can produce earthflows, lateral spreading and the loss of bearing capacity of soils underneath the foundation of buildings and other structures.

2.2.2 Earthquakes and tsunamis

Earthquakes occur throughout almost all the Archipelago. Figure 2.3 shows the location of epicenters during the last decades.

According to PHIVOLCS the Country is hit by an average of 5 earthquakes per day. Table 2.1 shows Intensity, Magnitude and number of casualties of the most destructive events during the last four decades (Earthquake and Tsunami, 1990).

Giant sea waves called tsunami (from the Japanese), mainly produced by submarine earthquakes, volcanic eruptions and huge submarine landslides, also occur along the coast of the Philippines, but are less frequent than earthquakes.

Twenty-seven known tsunamis affected the Philippine coasts from 1603 to 1975. On 16 August 1976 a tsunami, induced by the Moro Gulf earthquake, devastated the southwestern coast of Mindanao. The 5-meter high waves killed 3,000 persons, injured 8,000 and left about 12,000 families homeless (Earthquake and Tsunami, 1990).

Figure 2.4 shows tsunami-prone areas in the Philippines and their relationship with the tectonic setting of the Archipelago.

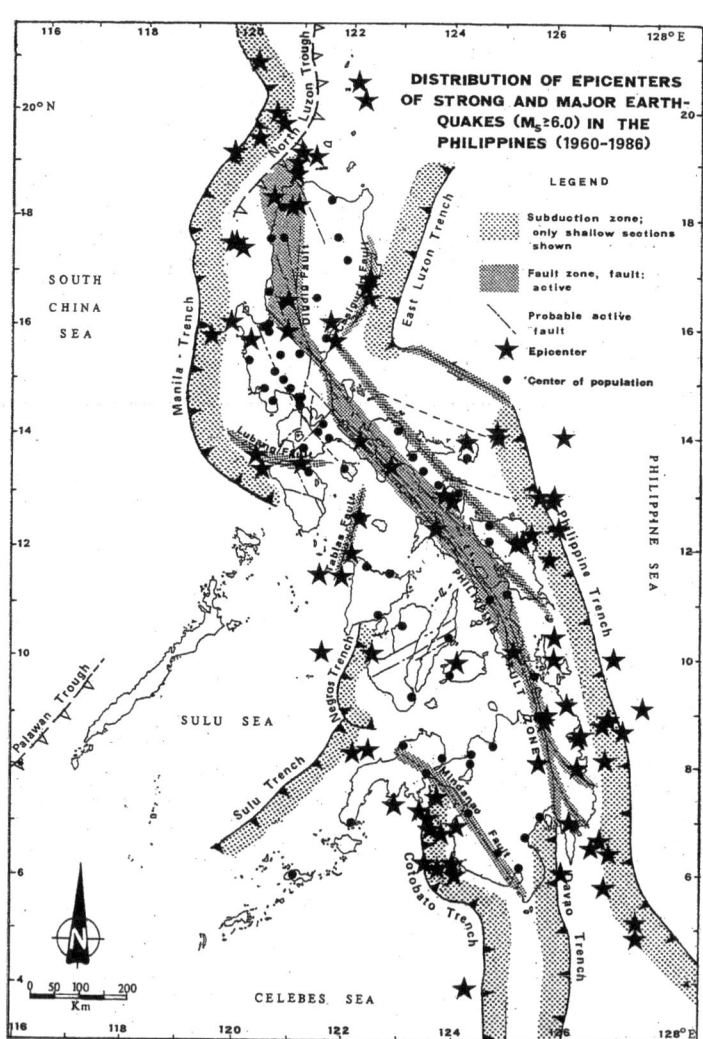

Fig. 2.3 – *Distribution of earthquake epicenters (1960-1986) within the Philippine Island Arc after Punongbayan (PHIVOLCS, 1987).*

24

2.2.3 Volcanic eruptions

The Philippines being located along the Pacific Ring of Fire suffer a remarkable amount of volcanic activity. Forty-one destructive eruptions occurred in the Archipelago during the period 1572-1991, an average of one major event about every 10 years. Intermittent minor eruptions, however, accompanied by smoke, tremors and heat are quite frequent in active volcanic districts. Table 2.2 gives details of major eruptions of the most dangerous volcanoes in the Philippines.

Volcanic eruptions are concentrated along some major tectonic lineaments. Out of 220 volcanoes, 21 are classified as active (Punongbayan, 1987), most of the others are considered to be inactive at present, while the remaining are dormant (Fig. 2.5).

An example of recurring volcanism in a highly populated area of the Philippines is the activity of Mt. Taal, 60 km south of Manila. According to Hargrove (1991) in 1754 a several month-long period of violent phreatomagmatic activity ended by partially modifying the volcano's shape, by closing up the wide channel connecting the caldera to the sea and changing it from a saltwater lagoon into a freshwater lake. The next violent event occurred in 1911 with an electrical display visible from a distance of hundreds of kilometers. Huge columns of smoke, mud, fine ejecta, earthquakes and widespread destruction were associated with the eruption. Metro Manila is built on volcanic ejecta and is crossed by an active tectonic lineament, the Marikina Fault. The most recent eruption in the Philippines was that of Mt. Mayon (Legaspi, southern Luzon), which resumed its activity on February 3, 1993. Recent previous eruptions occurred in 1984 and 1977.

2.2.4 Landslides and liquefaction

The widespread slope instability of the philippine mountains is mainly due to the interaction of geologic factors and seasonal rains. The landforms of the Archipelago are commonly affected by an accelerated evolution because of the continuous upheaval of the region, the related steepening of

TABLE 2.1 - Destructive earthquakes in the Philippines from 1954 to 1990					
Date	Epicenter	Intensity	Magnitude	Dead	Injured
July 16, 1990	Rizal (Nueva Ecija)	VII	7.7	1,666	3,561
Aug. 17, 1976	Moro Gulf (Mindanao)	VII	7.9	3,739	8,000
Apr. 7, 1970	Baler (Quezon)	VII	7.3	15	200
Aug. 2, 1968	Casiguran (Aurora)	VII	7.3	270	600
Apr. 1, 1955	Lanao (Mindanao)	VII	7.5	291	713
July 2, 1954	Bacon (Sorsogon)	VII	8.3	13	101

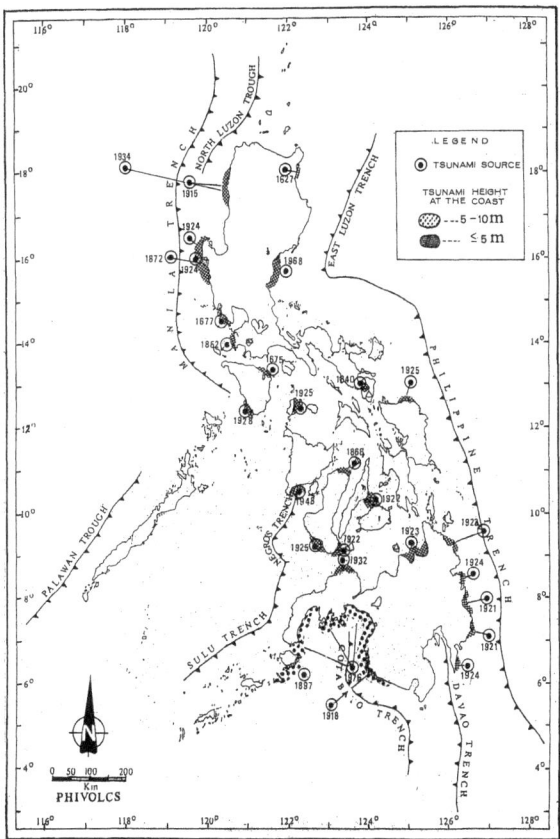

Fig. 2.4 – *Tsunami prone areas of the Philippines after Uy and Punsalan (PHIVOLCS, 1987).*

slopes, the rejuvenation of river valleys and the strong erosion induced by the high-intensity tropical rains. Earthquakes and volcanic eruptions in general powerfully contribute to shape the morphology. The former generate landslides through the shaking of slope materials, the latter by depositing huge quantities of pyroclastic products with the potential for sliding. The landscape of the Archipelago, moreover, is no longer protected and stabilized by the natural forest vegetation, which has been largely removed by the exploitation of forest resources and wildfires.

A variety of types of landslides are usually triggered by strong quakes in the mountainous and hilly provinces of the Philippines. Seasonal landslides, which are rather common in the Country, have exerted yearly recurring impacts on the landscape frequently damaging roads and other infrastructure during this century. The importance of mass movements in Luzon and the related damage were clear after the July 1990 earthquake, the 1991 eruption of Pinatubo and the monsoon rains which started soon after these events.

Mudflows (also termed lahars in Indonesia) are huge downward movements of pyroclastic material which accumulates on the slopes of volcanoes during eruptions. Heavy rains by causing downhill movement of finer ejecta trigger the initiation of mudflows. Once the lahar is in motion under the effect of prolonged rains it can receive and carry coarser sediments, blocks and boulders. When the fluid mass reaches the flat land it usually has devastating effects on infrastructure, agriculture and the environment. Sand layers from lahars related to past eruptions were discovered in various places on the Central Plain.

During earthquakes, water-saturated sand deposits can liquefy under the effect of ground shaking. This means that the soil becomes fluid with dangerous consequences for houses and other structures. When liquefaction occurs, foundations can sink, tilt or undergo differential settlement.

Based on Spanish chroniclers reports the area around Dagupan City (Lingayen Gulf) in the Central Plain is thought to have undergone liquefaction at least twice in historical times. Liquefaction occurred during the 1792 and 1896 quakes, as well as during the July 16, 1990 event. Before the July 1990 tremor, liquefaction had been triggered in Panay Island (Central Philippines) during the May 1990 quake which caused considerable damage (Observer, PHIVOLCS, 1990). Fossilized traces of sand liquefaction in various parts of the Philippines were recently discovered, indicating that this phenomenon is quite common throughout the Archipelago (IMAI et al., 1991).

2.3 Climate-related disasters

The vulnerability of the Philippines to climate-related disasters is mainly due to the many powerful typhoons yearly crossing the southwest Pacific Region. The high speed winds and abundant rain-

TABLE 2.2 - Major volcanic eruptions in the Philippines during the period 1572 - 1993 (adapted from ADB, 1991a)	
Names of Volcanoes	Eruptions
Pinatubo (Zambales)	2 eruptions, the most recent in June 1991; previous one about 400 years before.
Mayon (Legaspi)	44 significant eruptions since 1616. Worst eruption was in 1814 (over 1,200 casualties). The most recent eruptions occurred in 1993, 1984 and 1977.
Taal (Manila)	33 eruptions since 1572. The largest occurred in 1754 and 1911 (1,300 casualties).
Bulusan (Sorsogon)	First eruption in 1852, followed by intermittent eruptions, most recent in 1985.
Canlaon (Negros)	21 eruptions, first in 1866 and last in 1985.
Hibok-hibok (Northern Mindanao)	Activity recorded between 1948 and 1953 with most destructive phase in December 1951 when an avalanche killed more than 3,000 persons.

TABLE 2.3 - Number of typhoons, 1948-1990	
Year	Typhoons
1948-1986	189
1987	6
1988	4
1989	7
1990	8

falls, which recurrently hit the area, bear an enormous destructive potential. The devastation is magnified by the fact that human activities are often concentrated in flat, flood-prone areas.

The Philippines are located in the typhoon-prone area of the Pacific and crossed by 20 to 30 typhoons per year mainly in the period from June to November. The most abundant monsoon rainfall is usually concentrated in July, August and September. The number of destructive typhoons during the period 1948-1990 is given in Table 2.3.

Figure 2.6 shows typhoon tracks across the Philippines during the period 1955-1985. Ninety-five percent of the typhoons crossing the Philippines originate in the Pacific Ocean, while the remaining 5% come from the South China Sea. The eastern coast of the Archipelago is, thus, the most vulnerable, Northern Luzon having the highest frequency of typhoons during the year. Table 2.4 shows wind speed, number of casualties, wounded and dispersed people during 1986-91 typhoons. The destruction of houses, port facilities, drainage systems, agricultural land and infrastructure should be added to these figures.

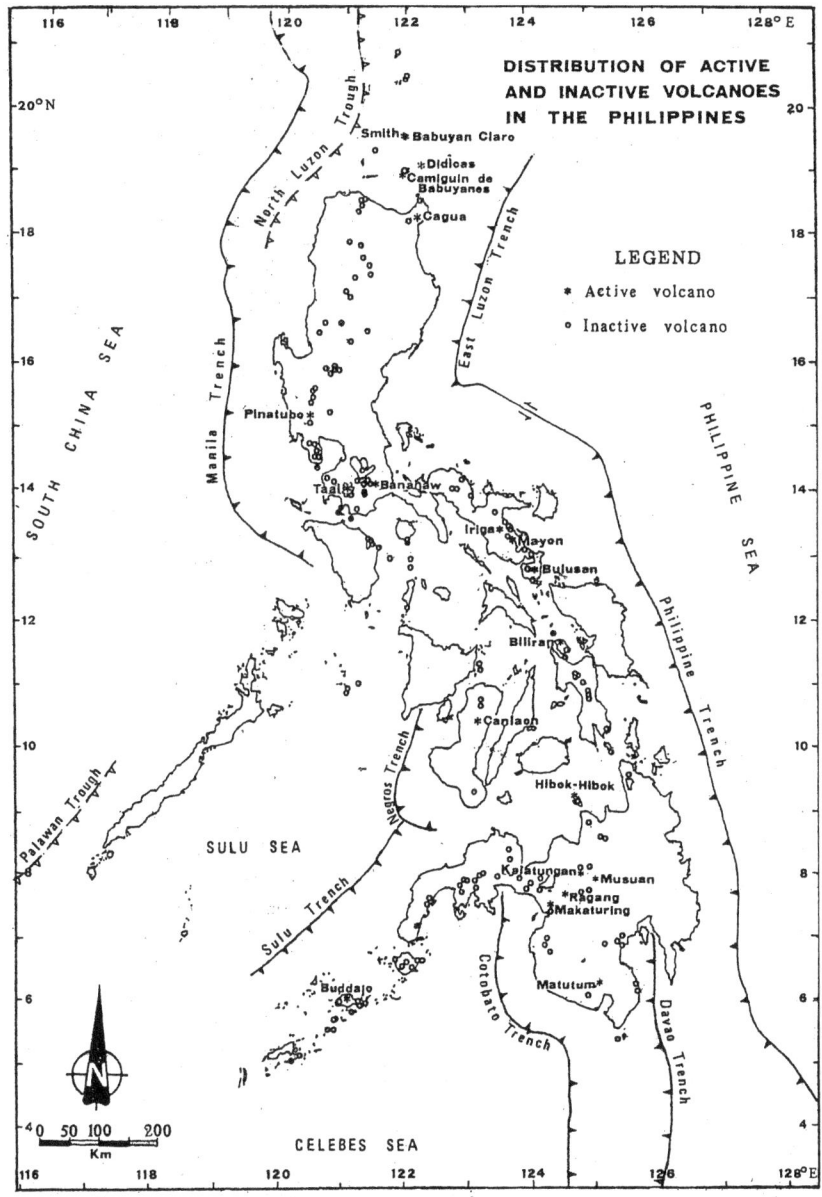

Fig. 2.5 – *Distribution of volcanoes in the Philippines after Punongbayan (PHIVOLCS, 1987).*

The exponential trend of the human losses shown in Table 2.4 may be partly due to a more reliable assessment of damage within the affected areas, partly to the inflation of losses and damage at the local level to attract government funds.

According to PAGASA, 47% of the annual average rainfall is related to typhoons. The low-lying lands, which represent about 40% of the total land surface of the Philippines, are particularly vulnerable to flooding. In July 1972, during a record rainfall, low-lying areas in Pampanga and Bulacan provinces (northwest and north of Manila, respectively) were submerged by over 2 m of water. Severe province-wide flooding occurred in 1974, 1978, 1981 and 1986. Almost every year, parts of Manila and the surrounding flat land are flooded. The highest rainfall concentration occurs when slow-moving or almost stationary typhoons follow one another, thus extending the period of heavy rainfall. In urban areas, the lack of maintenance of drainage facilities increases vulnerability to flooding. The primary effects of typhoons such as the destruction of houses, infrastructure and services are often followed by devastating secondary damage due to flooding and landslides.

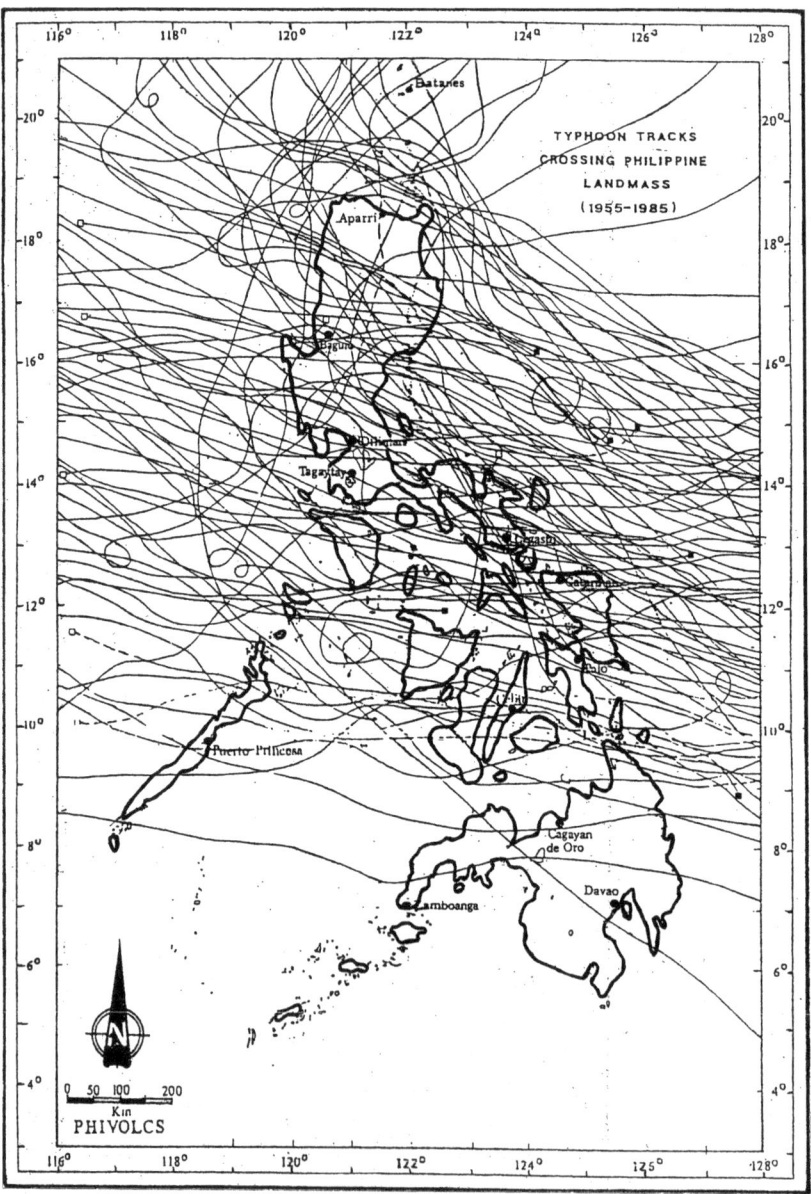

Fig. 2.6 – *Typhoon tracks over the Philippines, 1955-1985, (Nufable, 1986) after Punongbayan (PHIVOLCS, 1987).*

TABLE 2.4 - Data on typhoons in the Philippines, 1986 - 1991					
Typhoon Name	Date	Maximum Wind speed in km/h	Dead	Injured	Missing
Thelma	Nov. 1991	200	5000	2500	
Ruping	Nov. 1990	205	508	1274	240
Unsang	Oct. 1988	215	157	316	60
Sisang	Nov. 1987	240	979	927	
Pepang	Oct. 1987	220	141	67	
Herming	Aug. 1987	240	94	468	
Gading	Jul. 1986	205	89	16	20

2.4 The 1990-91 geological disasters in Luzon

The recent disasters which struck the Philippines during 1990 and 1991 added up to a regional calamity with loss of lives and unprecedented destruction (Fig. 2.7). The earthquake which hit the island of Luzon on July 16, 1990 had a Magnitude of 7.7 (Richter) and produced a spectacular major ground rupture over a distance of 120 km, plus minor surface faulting.

The unprecedented length of the major rupture, its 6.2 m of maximum horizontal displacement, the regional-scale liquefaction in Central Luzon, and the numerous landslides in the mountainous areas are among the most destructive and best documented geological phenomena of this century. The seismic swarm which soon followed the quake and lasted for a few months was interpreted as a subsurface rearrangement of basement blocks in the area.

The subcrustal reorganization was in turn responsible for triggering the intrusion of molten rock, and the 1991 awakening of Mount Pinatubo and Mount Taal, two volcanoes located north-west and south of Manila, respectively.

The powerful June 1991 eruption of Pinatubo spewed volcanic ash over five Central Luzon provinces, blanketing almost 10,000 square kilometers of landscape, and whitening South China Sea waters; the explosion caused major destruction and fatalities. The primary effects of the earthquake and the eruption produced huge environmental impacts, but the additional devastation of the landscape which followed these events was caused by the heavy monsoon rains of 1990-1994. For months abundant downpours initiated massive slope instability and catastrophic erosional processes.

The interaction between the rains and the soils loosened by the quake on the one hand, and the ash deposited by Pinatubo on the other, produced huge downslope movements that exacerbated the effects of the primary damage. A factor that facilitated the mobilization of sediments was the widespread deforestation in Luzon during the last decades.

Most Filipinos, who have experienced a number of calamities during their lifetime, have traditionally fought adverse natural phenomena. In a famous church in Intramuros (Manila), for example, an inscription states that it was demolished four times by earthquakes, badly damaged by typhoons and finally reconstructed after bombing during the Second World War.

A significant difference exists, however, between the disasters of 1990-1991 and those of the past. Until a few decades ago the Country was sparsely populated and more extensively covered by rainforest, thus, the damage was much smaller. The population mostly lived along the coastline and hazards such as landslides, flooding and lahars had a lesser impact. The 1990-91 disasters took place in a densely populated Country, in which large parts of the mountainous areas had been deforested.

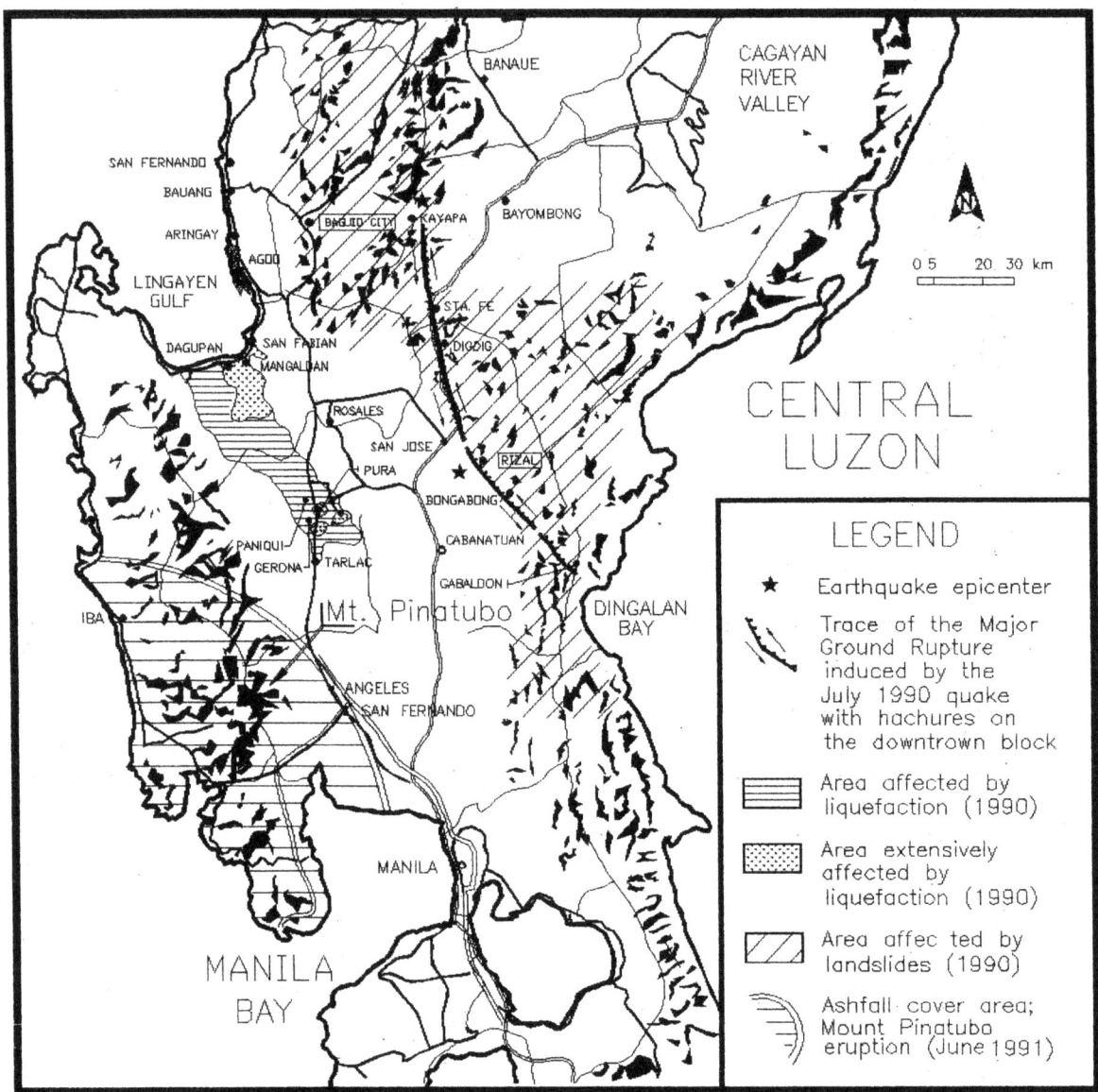

Fig. 2.7 – *Overview of the 1990 and 1991 disasters in Luzon (the area affected by ground ruptures, landslides and liquefaction was adapted from Punongbayan and Umbal, 1990). The ash blanket within about 50 km from the crater was deposited during the major explosive episode (June 12-15, 1991). Winds from the storms blew ash to Manila and further south as well as west towards Vietnam and Cambodia.*

Chapter 3

TECTONICS, SEISMICITY AND VOLCANISM OF LUZON

3.1 The Pacific Plate Region

The dynamic activity of the lithosphere is one of the most influential factors in the evolution of the surface of our planet. The earth's crust is composed of mobile plates of various shapes and sizes, which may be stationary for some time, or colliding, fracturing, spreading or diverging. Large plates are likely to be affected by some of these conditions at the same time.

The Pacific Plate (Fig. 3.1) is one of the most active, with high seismicity and significant volcanism around its periphery. About 80% of the world's active volcanoes as well as a considerable number of the yearly total of earthquakes are located along the Pacific Ring of Fire. The Plate is bordered by smaller crustal blocks: the Nazca and Coco Plates on the east, responsible for eastward subduction and high seismicity along the Chile-Peru-Central America Trench, and the northwest-ward-moving Philippine Sea Plate on the west, responsible for subduction beneath the Philippines (Fig. 3.2). The Archipelago, which is located at the convergence of the Eurasian and Philippine Sea Plates, forms part of a 4,000-km island arc stretching from the Kuril Basin in the north to Indonesia in the south.

According to Taylor and Hayes (1983) a phase of important crustal evolution affected the tectonic setting of the southwestern Pacific about 20 million years ago. A global plate rearrangement is thought to have taken place in that area during the middle to late Eocene (Appendix A), probably associated with the folding and uplifting of the Himalayan Belt. A fundamental contribution to the understanding of geodynamic processes in the western Pacific region and its present tectonic setting has been made by the studies conducted in recent years. Bathymetric variations, shallow and deep seismicity, vulcanism, seismic reflection profiles, gravimetric anomalies, surficial traces of faults and ocean floor geology have proved to be in good agreement with the plate tectonics scenario in the Philippines.

The double-sided reverse underthrusting of the ocean floor beneath the Country, the island-arc deformation, the associated high seismicity and volcanism are interrelated processes strongly affecting the environment of the Philippines. Due to the concentration of these phenomena in a relatively small part of the southwestern Pacific, the Archipelago is one of the planet's most disaster-prone areas.

3.2 Morpho-tectonic Units of Central and Northern Luzon

The Middle Miocene Philippine Fault, which was initially recognized more than a century ago, marks a strong physiographic contrast, separating the Central Luzon plain from mountainous northern Luzon. Six morpho-tectonic units can be recognized on the mid-northern part of the island (Fig. 3.3).

The N-S trending Cordillera Central, 300 km-long and 90 km wide, is the major tectonic unit; it runs along the western side of Luzon with maximum elevations of about 3,000 m. Acid plutonic rocks form the core of the chain, the outer shell of which consists of shallow- to deep-sea sedimentary rock formations with intercalated volcanics. The uplift of the Central Cordillera batholiths started dur-

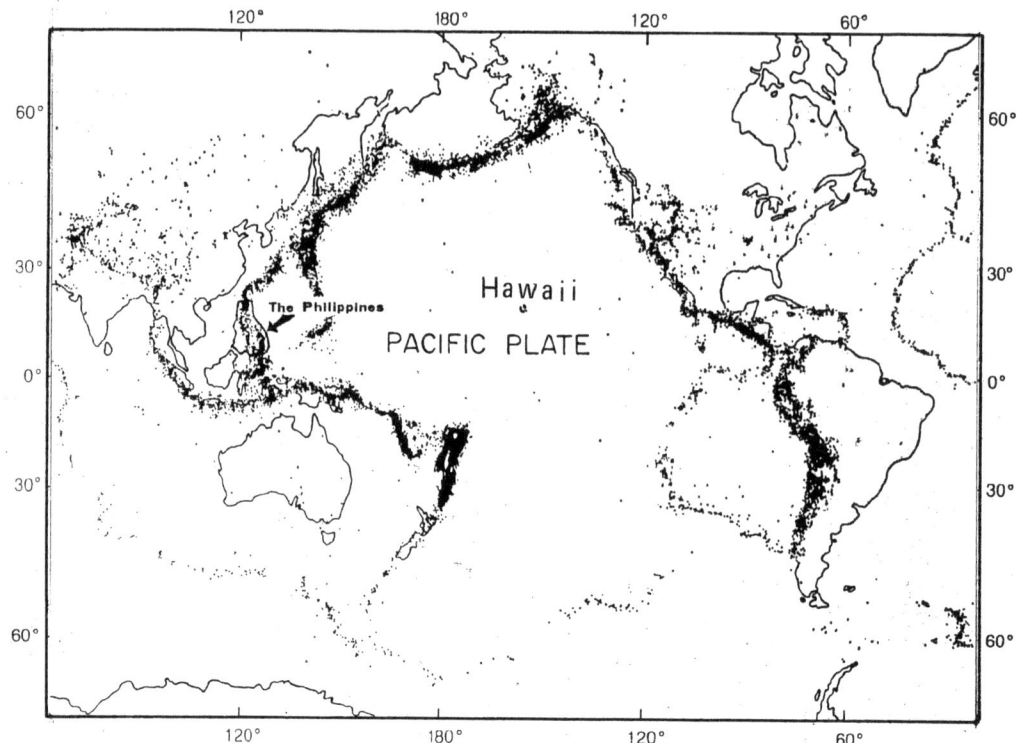

Fig. 3.1 – *Map showing earthquake distribution along the Pacific Plate margin during the period 1961-1967 (USGS). The periphery of the plate is also well known as the Pacific Ring of Fire due to the presence of numerous active volcanoes.*

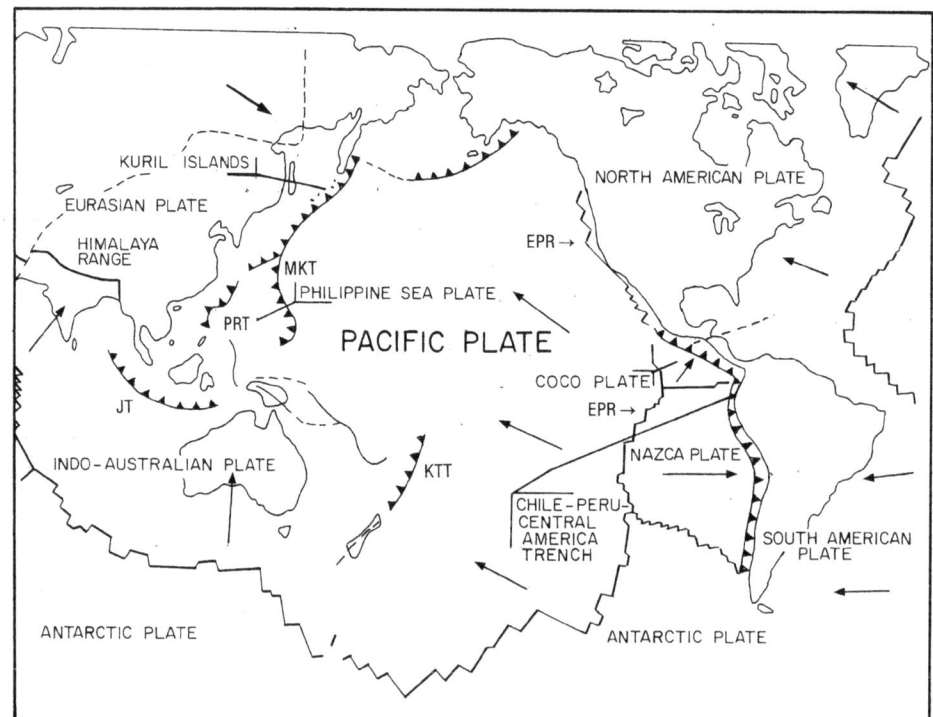

Fig. 3.2 – *Map of the Pacific Plate and sub-plates, with the Coco and Nazca Plates to the east and the Philippine Sea Plate to the west (MKT: Marianas-Kuril Trench; PRT: Philippine-Ryukyu Trench; KTT: Kermadec-Tonga Trench; JT: Java Trench, EPR: East Pacific Rise).*

ing the Miocene. The Sierra Madre, which borders the entire eastern side of the island, has an overall length of 600 km and maximum elevations of 1,500 m. This Range, which also consists of acid intrusive bodies, is divided into a northern and a southern segment by the Philippine Fault near Dingalan Bay.

The connection between the southern zone of the Cordillera and the Sierra Madre is marked by the presence of a third morpho-tectonic unit, the Caraballo Mountains. With a Pre-Tertiary basement made of schists and tonalites unconformably overlain by sedimentary, volcanic and pyroclastic rocks, the Caraballo Range is a smaller unit with comparatively lower elevations.

These three units form the catchment basin of the fourth unit in Luzon, the N-S oriented Cagayan River Valley, which is the second largest expanse of flatland in the Philippines. The fault-bounded Cagayan archipelagic basin, 200 km long and over 50 km wide, is almost completely surrounded by these mountains, except on the northern side, and mainly consists of Oligocene to Quaternary clastic sediments.

The 200 km-long, 80 km-wide Central Plain (fifth unit) stretching from the Lingayen Gulf to Manila Bay is a N-S oriented depression forming the largest area of flatlands in the Philippines. Bounded to the northeast by the Philippine Fault and to the west by the Zambales Range, the Central Plain depression was filled with loose clastic sediments during Tertiary and Quaternary times. According to Bachman et al. (1983) a sedimentary sequence up to 14 km thick was identified through multichannel seismic reflection in the center of the Plain. Surficial deposits of sand along riverbanks, due to lahars (mudflows) deposited during the Quaternary, were recognized after the 1991 Mt. Pinatubo eruption. Isolated Quaternary volcanoes interrupt the otherwise unrelieved monot-

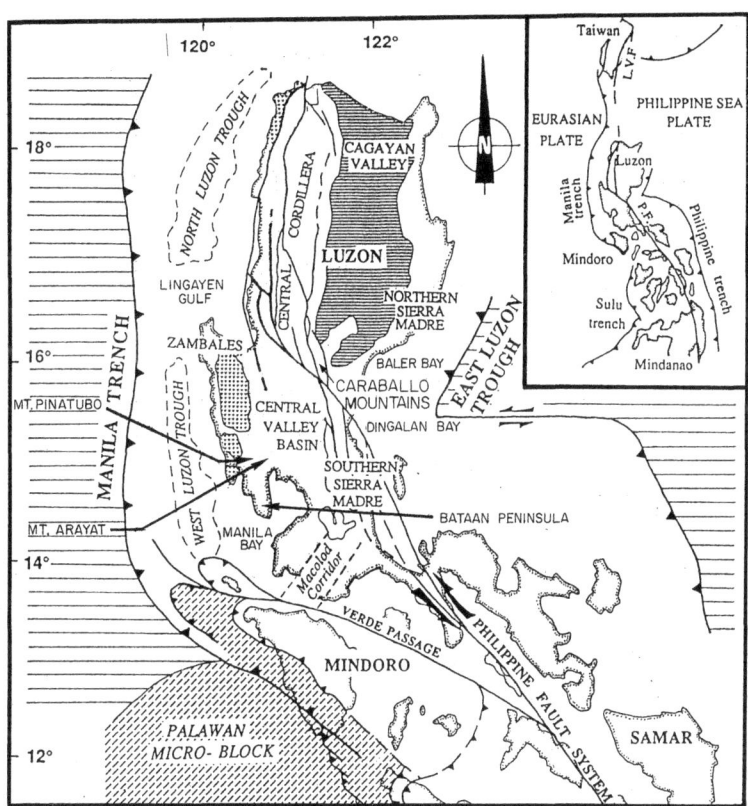

Fig. 3.3 – *Physiographic and tectonic features of Luzon Island (Pinet et al., 1990).*

ony of the plain; Mount Arayat (1,030 m), the most prominent item of relief midway between Manila and the Gulf of Lingayen, is one such example.

West of the Central Plain lies the Zambales Mountain Range, the sixth unit, which extends southwards from the Lingayen Gulf for about 150 km. The range is composed of Tertiary ophiolitic rocks and its southeastern end is in contact with a sequence of Quaternary volcanoes known as the Bataan Orogen. Of the 27 vents forming the lineament, the most famous is Mt. Pinatubo, with its peak rising to 1, 732 m before the collapse of the crater after the 1991 eruption (Chapter 8).

3.3 Tectonic setting of Luzon

3.3.1 Lineaments and bathymetry

Luzon's major tectonic lineaments are the double-sided subduction due to the convergence of the Eurasian and Philippine Sea Plates and the horizontal slip motion along the Philippine Fault. Most researchers consider the left-lateral strike-slip motion along this several hundred km-long lineament to be the mechanism which accommodates part of the crustal shortening induced by the reverse polarity underthrusting of the ocean floor beneath the island. Relevant to this dynamic framework (Lewis & Hayes, 1983) are major sea-floor topographic features of the island arc (Fig. 3.4):

a) the 5,000 to 6,000 m deep Manila and Philippine Trenches located to the west and east of the Archipelago, respectively;

b) the flat-floored depression known as the East Luzon Trough;

c) the massive, basaltic Benham Rise, a topographic high, with a depth range of 2000-3,000 m locally ascending to a few dozen meters below sea level at Benham Bank, and

d) the South China Sea Plate west of the Archipelago.

The shape of the Philippine Island Arc has most probably been influenced by the presence of Benham Rise, which is a thickened portion of the Philippine Sea Plate oceanic crust. According to Ringenbach et al. (1991) the similarity of the shape of Benham Rise to the sharp bend in the Luzon coastline suggests that the resistance of the basaltic sea floor to subduction may be the reason for the bending of the Philippine Fault and the tangential shape of its splays.

3.3.2 Tectonics

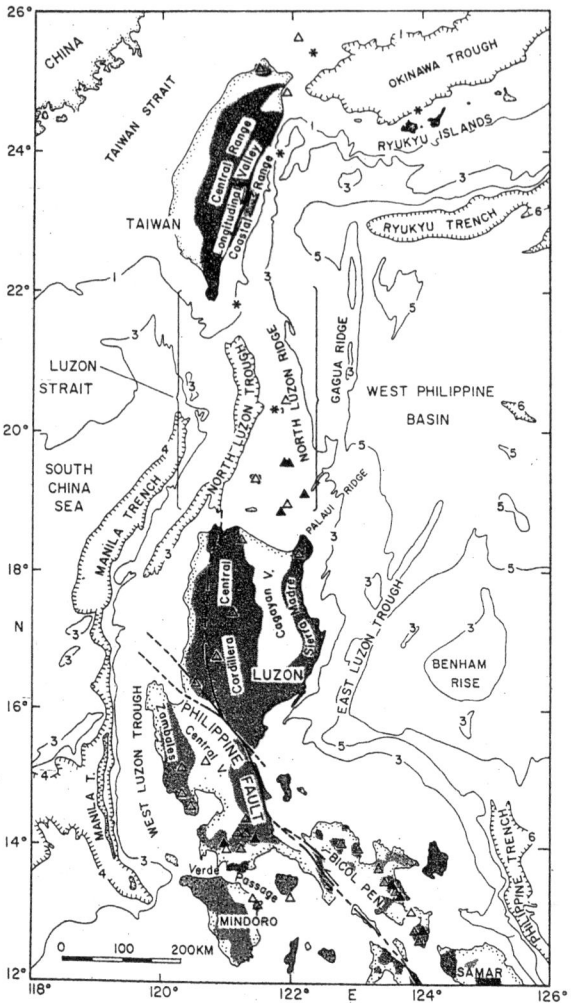

Fig. 3.4 – *Bathymetry of the Philippine Archipelago in thousands of meters (Hamburger et al., 1983). Reprinted by permisssion of the American Geophysical Union.*

Convergence of plates, the eastward subduction of the South China Sea Plate along the Manila Trench and the westward subduction of the Philippine Sea Plate along the Philippine-East Luzon Trench and, finally, the Philippine Fault are the most important tectonic elements of the Archipelago (Fig. 3.3). The Philippine Fault, a 1,300 km-long active lineament with left-lateral strike-slip motion, has played an important role since the start of the subduction.

With its seismicity, marked ground rupture features and evident morphologic contrasts, this fault, through a considerable regional deformation, absorbs part of the crustal shortening induced by the converging plates. The complex dynamics due to opposite trending subductions and the horizontal-slip along the Philippine Fault, subject subsurface stress fields to a continuous change in equilibrium, which results in countless minor tremors and frequent medium to strong earthquakes.

At the southern margin of the Caraballo Mountains, near Rizal City, the Philippine Fault splits into a number of northwestern then northern trending splays (Fig. 3.5), with a typical horse-tail pattern.

Due to this branchlike pattern of faults, the Central Cordillera is divided into a number of N-S trending structural units limited by the Coastal Thrust, the Pugo Fault, the Tuba and Abra River

Faults and the Digdig Fault. Philippine Fault splays are also interconnected by some transverse faults as for example the San Juan-Tebbo Lineament.

According to Pinet et al. (1990) and Ringenbach et al. (1992) the first Middle-Miocene movements along the Philippine Fault in Western Luzon were accommodated by strike-slip motion along the Abra River Fault. The Digdig Fault, which is connected to the Abra Fault by a NW trending segment near Baguio (Fig. 3.5), played a major role in the July 1990 earthquake.

In general, the division of the Philippine Fault into a number of subparallel lineaments suggests that strike-slip motion might have been channeled through different splays during Luzon's tectonic evolution. The core of the Cordillera Central (made of deep rooted batholiths) most probably reacted rigidly to the movement of the different blocks in the area, thus generating the splays as a sort of multiple-choice solution to the complex stress field.

The intricate fault pattern, which is well identified in Baguio-Lingayen Gulf region, is responsible for the local tectonic style and the destructive effects of the July 1990 and previous strong earthquakes.

Slip vectors of the Philippine Sea Plate region (Fig. 3.6) are NW oriented, while the motion of the South China Sea Plate is considered to be N to NE oriented.

Barrier et al. (1991) evaluate the northwestward displacement of the Philippine Sea Plate at 7.4 cm/year, while a slower rate of 2-3 cm/year is proposed by other researchers for the NE motion of the South China Sea plate. Since for the Philippine Fault a horizontal displacement rate of 2-3 cm/year and an age of 2 to 4 million years are proposed, the overall displacement is estimated at between 40 and 120 km.

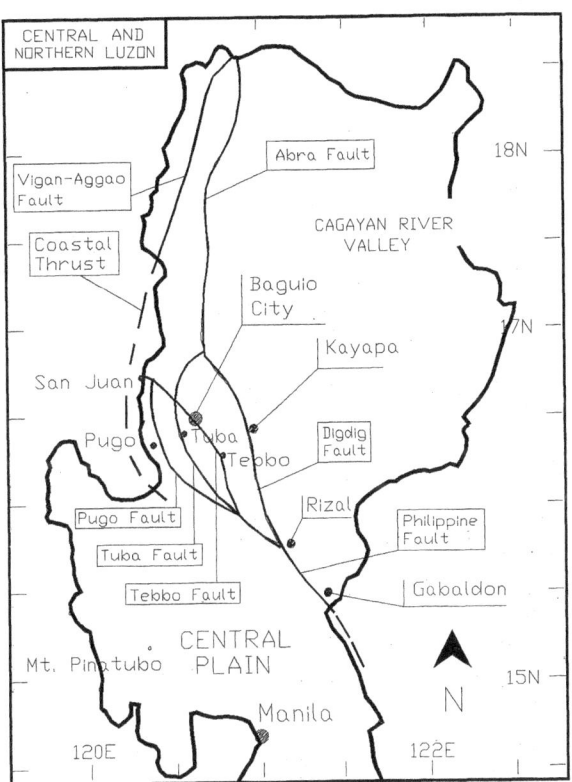

Fig. 3.5 – *Philippine Fault and Splays in Luzon.*

According to Seno (personal communication, October 1991, cited by Ringenbach, 1992) the crustal shortening, due to the convergence of Philippine and Eurasian plates, varies from 9.2 cm/year in Mindanao to 8 cm/year in Luzon. This considerable crustal reduction is thought to be partly compensated by the horizontal slip motion along the Philippine Fault and its splays, and partly converted into the uplift of the Central Cordillera.

By comparison, western and eastern parts of the Mid-Atlantic Ridge move apart at a speed of about one centimeter per year (Bonatti, 1994).

The uplift of the Central Cordillera is evident from the marked contrasts in elevation, youthful landform features, deeply incised steep-sided valleys, downcutting of streams and very active erosion. The rate of regional uplift during the Pleistocene is estimated at 1.5 mm/year by Ringenbach (1992), while the velocity of the convergence between the Philippine and Eurasian plates at Luzon's level is generally thought to be some 8 cm/year. About a quarter of it (2 cm/year or 2 m every 100 years) is compensated by the motion along the Philippine Fault during large earthquakes. Thus the July 16, 1990 quake with its maximum horizontal displacement of 6.2 m may have compensated about 310 years of cumulative crustal shortening. This timespan is quite close to the time elapsed since the previous strong earthquake which shook the same area in 1645 (SEASEE catalogue, 1985), i.e. 345 years before the Luzon quake (Fig. 4.9).

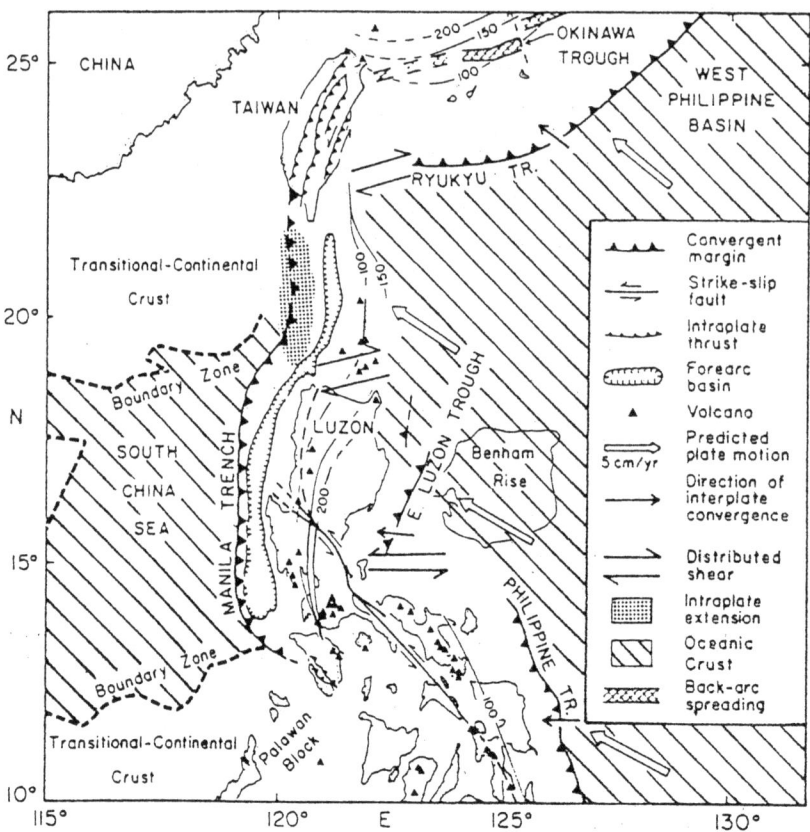

Fig. 3.6 – *Tectonic framework of the northern Philippine Island Arc (Hamburger et al., 1983) taken from Seno (1977); continent-ocean boundary zone from Taylor (1982); Okinawa Through from Lee et al. (1980). Reprinted by permission of the American Geophysical Union.*

3.4 Geologic history

The description of Luzon geology (Fig. 3.7) is limited to the northern and central part of the island, since the remaining areas were only marginally affected by the events described in this book.

The beginning of the orogenesis, Late Eocene - Late Oligocene, was marked by the uplift of granodioritic bodies forming the core of the Cordillera, which was still beneath the sea at that time, though there were some emerging volcanic islands. Due to bathymetric variations of the sea-floor, deep and shallow marine sediments were deposited. The sequence rests unconformably on an ophiolitic Cretaceous - Middle Miocene basement, made of basalts, schists and cherts. The Sierra Madre Range, which is similarly made of granodioritic bodies, is considered to be of Eocene-Oligocene age.

During Late Oligocene to Early Miocene times the major uplift and the emergence of the Cordillera batholith occurred, accompanied by shallow-water sedimentation on its flanks and voluminous volcanic activity. Fracturing, folding and faulting of sediments forming the outer cover of batholiths is associated with this highly dynamic stage. The metamorphosed and intensely deformed rock formations, intruded by the pluton, originally formed the shell within which the magma slowly cooled and crystallized. During the orogenic stage this shattered shell became the ideal structure for future porphyry copper mineralizations.

The tectonic activity continued in Middle Miocene to Holocene times while conglomerates, sandstones and pelites with interbedded volcanics were deposited. Surrounded from west to south and east by the Cordillera Central, the Caraballo Mountains and the Sierra Madre lies the Cagayan River Val-

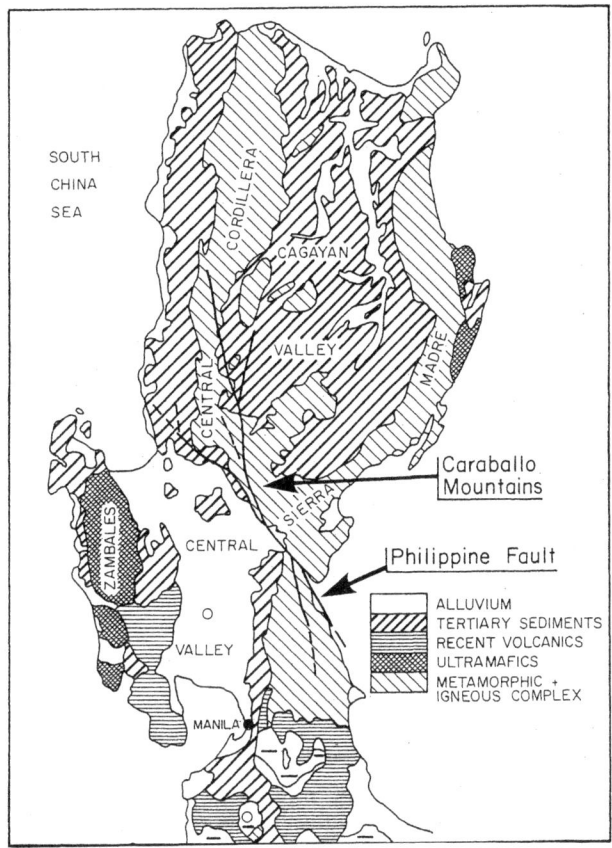

Fig. 3.7 – *Geologic reference map of Luzon and the location of the major morpho-tectonic units: Central Cordillera, Sierra Madre, Cagayan Valley, Caraballo Mountains, Central Valley and Zambales Range (Geary et al., 1983). Reprinted by permission of the American Geophysical Union.*

ley, a tectonic depression with a pre-Oligocene metamorphosed volcanic basement, filled with clastic sediments between the Oligocene and the Quaternary. The Caraballo Mountains, which rest on a pre-Cretaceous basement of schists and tonalites mainly consist of pyroclastic, volcanic and sedimentary rock formations ranging in age from Cretaceous to Eocene.

The Philippine Fault separates this part of Luzon from the Central Valley, a flat floored plain which was filled with loose sediments during the Tertiary and the Quaternary. Lastly, the Zambales Range, the most westerly unit in Central Luzon, is composed of a north trending assemblage of mafic-ultramafic rocks and basaltic to rhyodacitic volcanics. The emplacement of the Zambales ophiolite (Geary and Kay, 1983) took place during Middle to Late Oligocene times and is considered to mark the beginning of subduction along the Manila Trench.

3.5 Seismicity

Due to the active motion of plates in the archipelago numerous earthquakes are generated thus making the Philippines an area with a marked seismic hazard. The regional pattern of the seismicity shows a striking relationship with subduction dynamics, the inland and marine fault systems, major morpho-tectonic units in Luzon, and geophysical data. Clustering of seismicity, depth of hypocenters and focal mechanisms are also in agreement with this scenario.

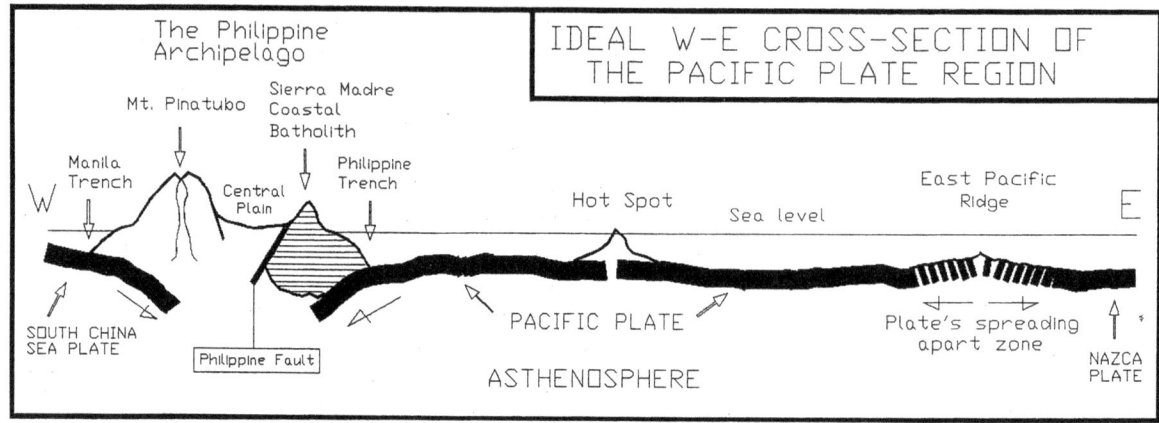

Fig. 3.8 – *Idealized cross-section of the Pacific Plate from the Philippine Archipelago to Nazca Plate, showing the three types of volcanism: subduction zones (Pinatubo), hot spots (Hawaii) and oceanic ridges (East Pacific Rise).*

Numerous and often large earthquakes (SEASEE catalogue, 1985) hit the Philippines during the past four hundred years. The geologic history of Luzon suggests, however, that seismicity must have been consistent also during the most critical phases of the uplift of the Cordillera and certainly during the Quaternary.

According to Su (1988), 18 earthquake generators (source zones) are responsible for the high seismicity of the Philippines (Chapter 4). An important aspect of the converging motion of plates is the role of Benioff zones since numerous earthquake foci cluster along these dipping planes extending to a depth of over 200 km (Hamburger et al., 1983). A number of major earthquakes in the Archipelago are also located along the Philippine Fault.

3.6 Volcanism

3.6.1 Volcanic activity within the Pacific Plate

The Pacific Plate is well known for being surrounded by a Ring of Fire, which includes most of the world's active volcanoes; some activity, however, takes place within the plate as well. In a broad sense two basic conditions can occur: where the plate is thinned or is already spreading apart, hot basaltic magma quietly rises from the asthenosphere; in contrast, plate collision or underthrusting of one plate under another is generally associated with acid and explosive volcanism.

The Pacific Plate Region can be broadly divided into two parts by a line connecting the Kamchatka Peninsula to New Zealand through the Hawaii. The eastern portion (Figs. 3.1 and 3.2) is an unvaried huge depression about 5,000 meters deep, bounded to the E by the East-Pacific Ridge and marked by numerous parallel fractures. Along the thousands of kilometers of ridge the plate's margins diverge and crustal spreading takes place with the creation of new lithosphere as a consequence of the rise of molten basalt.

The western portion of the Plate, between Hawaii and the Philippines, is marked by submerged ridges, trenches, subduction zones and deep flat-floored areas. These morpho-structural variations of the ocean floor are indicative both of more complex tectonics in the Western Pacific Region and of the intricate dynamics of the asthenosphere underneath. Examples of this greater complexity are the basaltic magma being quietly extruded by volcanoes over hot spots along the Hawaiian Ridge, the explosive volcanism along the western margin of the Plate and the subduction of the ocean floor along the Marianas-Japan-Kuril Trench, the Philippine and Ryukyu Trenches and the Kermadek-Tonga Trench (Fig. 3.2).

Three types of volcanism (Fig. 3.8) are associated with the Pacific Plate dynamics: mantle plumes or hot spots (Hawaii); creation of new lithosphere where plates spread apart (East Pacific Ridge); and explosive activity where plates converge (Pacific and Eurasia collision zone).

In the first two cases non-explosive behavior and fluid basaltic magma characterize the activity. The volcanism along the Hawaiian Ridge occurs as mantle plumes or hot spots and is attributable to the presence of vents over plumes of heat rising through the upper mantle. Mostly thermal energy is released through this type of volcanism. A similar molten-rock rise, but related to a different tectonic setting occurs along the East Pacific Ridge, where the plate is separating after breaking apart. A quiet non-explosive upwelling of basaltic lavas from the asthenosphere takes place over several thousand kilometers along this fracture. This type corresponds to volcanism along mid-oceanic ridges and contributes greatly to the formation of new lithosphere.

By contrast, the third type of volcanism, widely represented in the Philippines, is characterized by explosive activity and occurs along converging plate boundaries. Along collision zones numerous earthquakes cluster and also Ocean floor is consumed through subduction and melting. As a result of this process the acid magma, which is generated at some depth, can intrude through the fissured zone into magmatic chambers near the surface during critical plate-motion episodes and produce the explosive type of volcanism. Most of the energy in this case is released as explosive kinetic energy.

In general a very different geological environment and diverse dynamics mark the evolution along the Pacific Plate margins, compared to other plate zones. Activity along plate boundary and subduction zones is mainly explosive, since the composition of the magma is granitic and thus has a degree of viscosity which favors explosive behavior. Besides, the rate of rise of viscous acid magmas is lower, compared to basaltic lavas, and so this type of molten rock can block chimney channels more easily. Once this happens gas pressure increases, the system reaches a critical condition and an explosive event takes place. The trigger of the explosion is often the compression of the magmatic chamber and underlying fractures, a consequence of the crustal movements with which earthquakes are associated. Explosive eruptions, however, can also take place in a mantle plume type of volcano if seawater penetrates the volcanic chamber, thus inducing a phreatomagmatic explosion, such as, for instance, the 1924 Kilauea eruption in Hawaii.

3.6.2 The volcanic environment of Mount Pinatubo

The explosive eruption of Mt. Pinatubo in 1991 belongs to the third volcanogenic environment: the volcano is located between the Manila Trench and the Philippine Fault (about 90 km west of it) which ruptured in July 1990. The Luzon earthquake was indirectly responsible for the awakening of the dormant Pinatubo: the actual triggering factor of the June 12, 1991 eruption was the subsurface block rearrangement which started after the 1990 quake and went on for months. The long reassemblage stage which took place during the aftershock period, modified the pre-existing equilibrium and affected the rock basement of the Central Valley, thus favoring magmatic injections into the Mt. Pinatubo chamber and the consequent awakening of the volcano. Mount Pinatubo is part of a sequence of vents parallel to the Manila Trench subduction zone and named the Bataan Lineament (Wolfe and Self, 1983).

The active and dangerous Taal Volcano, 60 km south of Manila, also warmed up during the beginning of 1991, as a result of the Luzon earthquake, and the central island in the caldera lake was later evacuated. Most likely due to the greater distance from the ground rupture zone and the related attenuation of subsurface movements in the area, the Taal volcanic chamber was marginally affected. After a few months of warming up and smoke emissions the danger gradually subsided. It is worthwhile to recall that in historic times ash from Taal eruptions reached Manila causing huge destruction (Hargrove, 1991). Recent news from PHIVOLCS confirm that Taal resumed activity again during 1994.

CHAPTER 4

THE JULY 16, 1990 LUZON EARTHQUAKE

4.1 Overview of the earthquake and its aftermath

On July 16, 1990 a catastrophic earthquake of 7.7 Magnitude with epicenter near Rizal City hit Luzon causing 1,666 casualties and extensive damage (Fig. 4.1). The event was interpreted by PHIVOLCS as a multiple-shock quake with two major epicenters along the Philippine and Digdig Faults. A 120 km-long major ground rupture between Gabaldon in Dingalan Bay and Kayapa (30 km E of Baguio) and two minor rupture zones NW of Rizal City were associated with the ground-shaking. Along the major surface faulting the ground underwent a horizontal left lateral strike-slip motion, with maximum displacement of 6.2 meters, unprecedented this century (Sharp, USGS-PHIVOLCS Report, 1990).

Death and devastation induced by the tremors were accompanied by considerable damage in Central and Northern Luzon. Buildings and infrastructure facilities collapsed in Baguio City, in the coastal

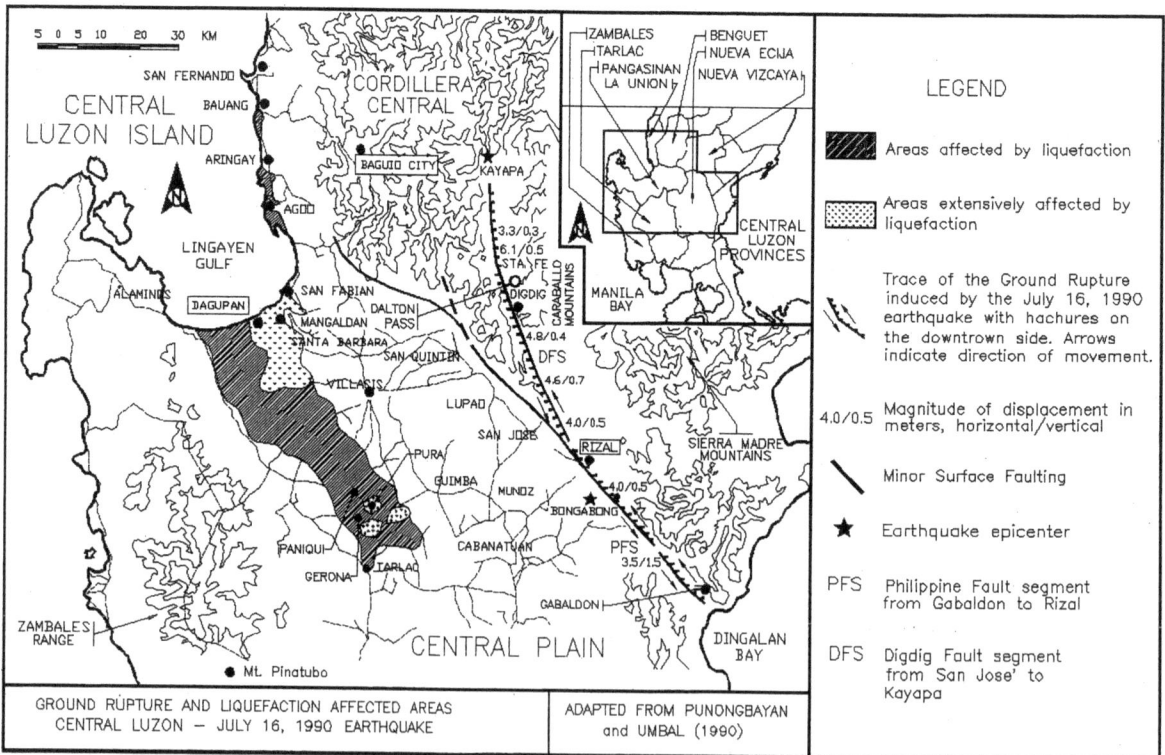

Fig. 4.1 – *Overview of Luzon 1990 earthquake area, adapted from Punongbayan and Umbal (1990).*

zone between Agoo and San Fernando in La Union (Lingayen Gulf) and in many localities in the Central Plain (Fig. 4.1). The entire zone near and along the ruptured fault segments was rocked the same day by further tremors.

Secondary short-range effects of the quake included landslides and liquefaction, both on a regional scale, while long-range effects involved a widespread sub-crustal block rearrangement in Northern and Central Luzon where two dormant volcanoes awakened during 1991.

The quake had an impressive and extensive environmental impact, triggering countless mainly shallow-seated landslides in the Cordillera Central and Caraballo Mountains. The slope materials mobilized by the tremor caused almost complete disruption of the transportation network in the region. Roads, highways and other communication systems in hilly and mountainous areas of western Luzon were severely damaged, thus blocking the immediate rescue efforts. From Metro Manila up to Abra province (Northwestern Luzon) there were signs of devastation. The impact on natural slopes can be regarded as one of Luzon's major disasters and as an example of landform evolution highly accelerated by the interaction of ground shaking and seasonal downpours.

Liquefaction occurred in a 70-km long and 20-km wide strip in Pangasinan and Tarlac provinces (Fig. 4.1). Sand boils, fountains and fissures were reported in the coastal areas of San Fernando in La Union province and in numerous other locations SW and W of the ground rupture in the Central Valley. Dagupan City on the Gulf of Lingayen was catastrophically damaged, while towns and villages in the Central Plain were badly affected.

After the major events of July 16, the region was rocked by further tremors and the ground continued to shake for months. The aftershock swarm was interpreted in terms of subcrustal block readjustment of Western and Central Luzon and mainly involved the Cordillera Central Mountains and the Central Plain rock basement.

For months a large number of tremors struck Baguio province where the fault pattern is comparatively more intricate. The final result of the underground rearrangement was the compression of shallow volcanic chambers and subsurface fractures some tens of kilometers west of the ruptured fault segment. Taal volcano (60 km SW of Manila) resumed activity in May 91 with the warming up of the central cone and smoke emissions. Mount Pinatubo (110 km NW of Manila) exploded in June 1991 with the widespread deposition of ejecta, after the volcano had been dormant for about 400 years. As a consequence of the eruption a region-wide blanket of light gray ash covered volcanic slopes and surrounding flatlands in the Central Plain. Then the transformation of the ash mantle into a fluid mass by heavy rains induced devastating lahars (mudflows) during 1991 and following monsoon seasons.

From July 1990 to October 1991 Central and Northern Luzon provinces were, thus, hit by interrelated disasters. The sequence of these catastrophic events was responsible for some thousands of deaths while a greater number of people were injured.

Over two million local inhabitants suffered, becoming homeless or jobless as a result of the damage to buildings, infrastructure and local activities. The agricultural potential of the region and the physical environment were drastically affected.

The damage to the economy of the Philippines by the quake, the eruption and their interaction with seasonal rains has been estimated at several billion US$, but long-term effects will be felt for years to come.

4.2 Geotectonic framework of Luzon

Out of the six morpho-tectonic units recognized in Luzon (Chapter 3), the Central Cordillera, Caraballo Range and Central Plain were most directly involved during the quake, but the major role was played by the Philippine Fault.

A 120-km segment, including part of the Philippine and the entire Digdig lineament, broke at the ground surface between Gabaldon and Kayapa during the July 1990 earthquake. The rupture, however, is considered to have continued underground for a further 50-100 km along the Abra Fault, which is presently considered as the northern extension of the Digdig Fault (Fig. 3.5). Although the importance of the Philippine Fault has been known since the end of the last century, its direction of motion was long a matter of controversy. The actual left-lateral strike slip motion was definitely proved by Allen in 1962 and is considered by Ringenbach et al. (1991) to result from the tangential component of the oblique convergence between the Eurasian and Philippine Sea plates. According to these workers a complex evolutionary process took place in Northern Luzon during the Miocene-Quaternary.

Left lateral displacement in the Middle Miocene-Pleistocene period occurred along the Philippine Fault which at that time presumably comprised several parallel N20W trending faults. During the Pleistocene the fault system was warped in Central Luzon, thus finally resulting in the present setting: the Philippine Fault with N140E strike and the N10-20E oriented splays with a typical tangential shape. Bending and subsequent horse-tailing attitude of the splays are attributed to the Middle to Late Eocene emplacement of the basaltic Benham Rise, which culminates 3,000 m above the surrounding depressions in the west Philippine Sea Basin (Lewis et al., 1983). According to Pinet and Stephan (1989) the initiation of the Philippine Fault took place in the Middle Miocene and the first strike-slip motion began in Upper-Middle Miocene times.

4.3 Seismicity levels within the Pacific Region and the Philippines

Before describing the Luzon earthquake, this section presents an overview of the trend of seismicity levels in the Pacific Region during the last 30 years and of the historical seismicity data for the Philippines. The earthquakes in the Pacific Region mainly occur along the Plate margins and intra-plate events are very few.

Figure 4.2 (upper sketch) shows the number of earthquakes per year (Magnitudes above 4) from 1964 to 1991 within the region including the Pacific Plate and adjacent smaller plates. In each column the lower part represents events on the Asian side, the upper part events on the American side.

The Meridian 150W has been used to separate the seismic events along the eastern and western boundaries of the Pacific Plate.

The position of this Meridian in the Gulf of Alaska approximately coincides with the northernmost zone of the East Pacific Ridge (Fig. 3.2), which is the major tectonic lineament of the region and divides the Plate into a western portion moving towards Asia (7-8 cm/year) and an eastern zone moving towards the Americas. The lower sketch in Figure 4.2 helps to visualize the boundaries of the zones involved.

The whole comumns (events within the entire Pacific Region) show a decreasing trend in the period 1965-77, with a minimun of 3,820 quakes in 1977; a sizable increase in seismicity occurs from 1978 onward, with a maximum of 6,290 quakes in 1986 and an average of 6,085 events per year in the period 1985-91.

In general, the seismicity along the coast of the Americas (upper part of the columns) decreases from about 1,500-1,600 to 900 events per year during 1964-1978 (with a minimum of 817 events in 1977); it is nearly stable within the narrow range of 1,124-1,175 quakes from 1979 to 1986, rising to 1,500-2,000 events in the 1987-91 period.

By comparison, the seismicity level along the western margin of the Plate (lower part of the columns) is mainly in the range of 3,200-3,300 events per year during 1964-1979, increases from 3,370 events in 1979 to 4,551 in 1985, and fluctuates in the range 4,090-4,540 events during the 1987-91 period.

Considering the overall seismicity of the Pacific region as broadly derived from the sum of earthquakes occurring along the western and the eastern boundaries of the plate, the Asian side appears to have been affected by a sizable increase in seismicity, compared to the American side, during 1977-86. The contribution to the western Pacific from the Aleutian Island Arc is not significant since the number of events per year, between 1964 and 1991, has a little variation over that period in the Aleutian Basin.

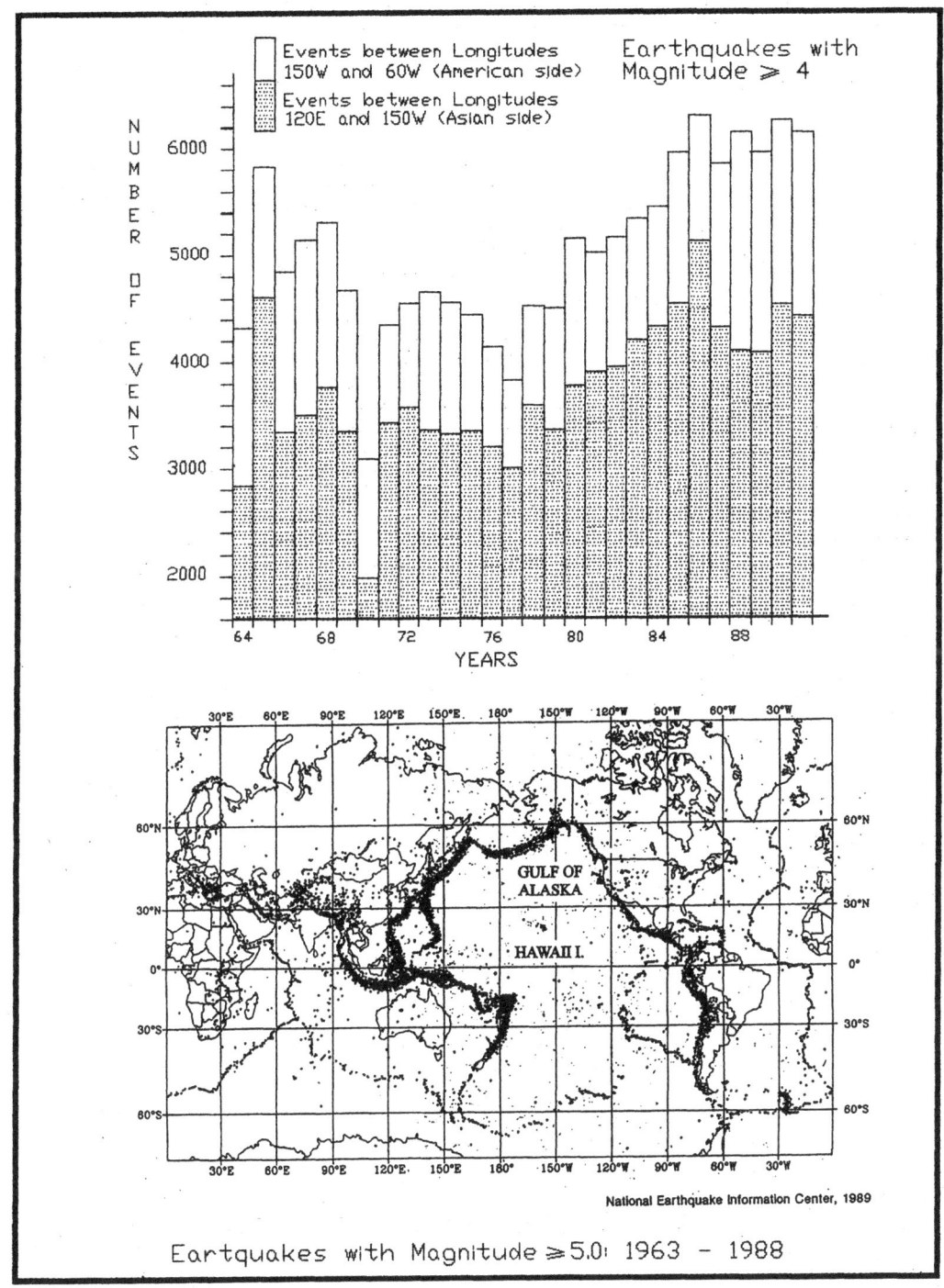

Fig. 4.2 – *The histogram (top) shows the number of earthquakes per year concerning the entire Pacific Region (whole columns), the western part of the Pacific Plate (lower columns) and the eastern part of the Plate (difference in height between the two), during the 1964-1991 period. The number of yearly earthquakes (above Magnitude 4) was derived from the catalogue of the International Seismological Centre (U.K.). The lower sketch (USGS-NEIC) visualizes the Pacific region and the areas to which histogram is related, that is Longitudes 120E - 60W (Lat. 75S-75N) for the entire Pacific Plate, Longitudes 120E - 150W for its western portion and Longitudes 150W - 60W for the eastern part.*

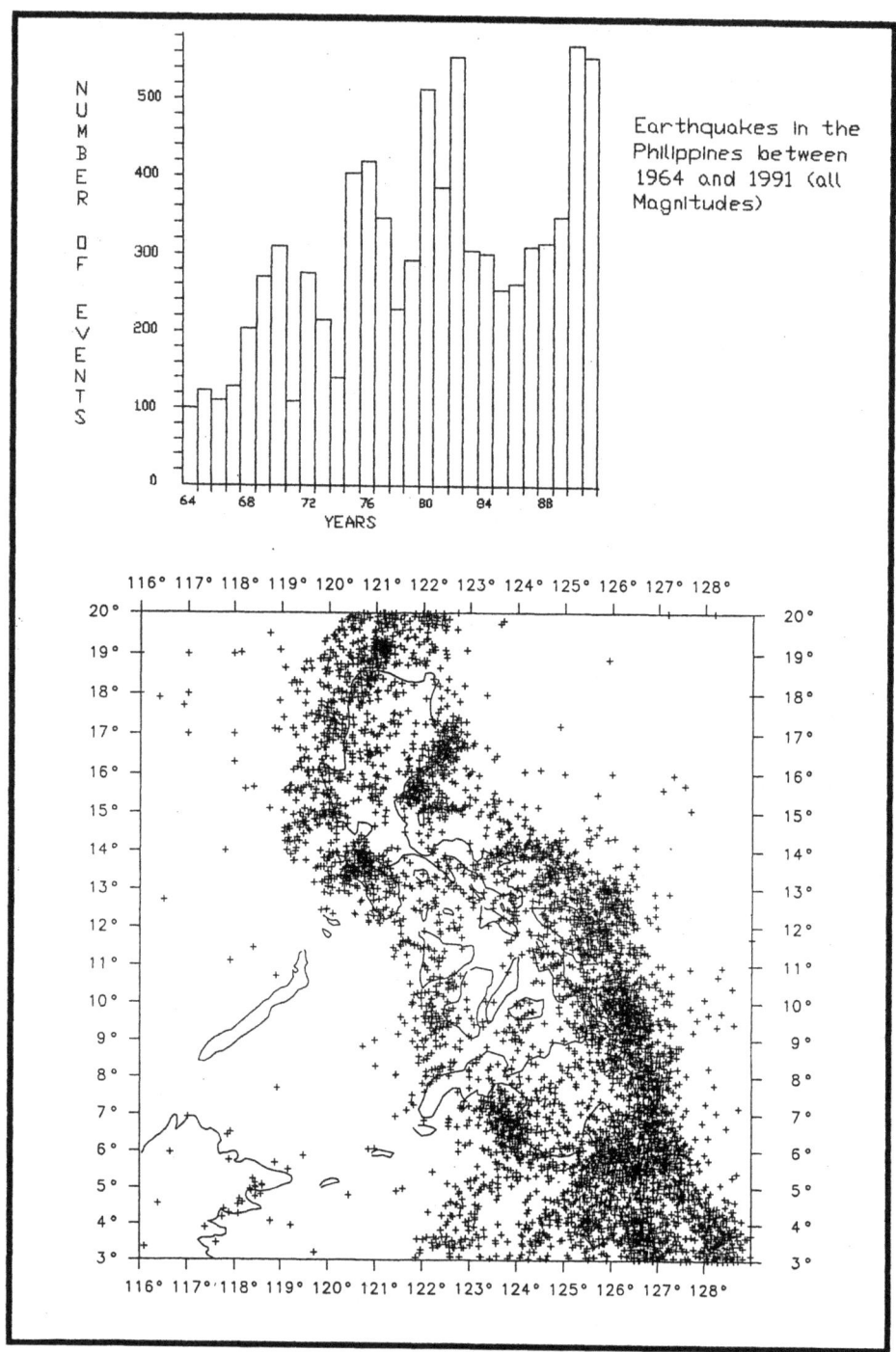

Fig. 4.3 – *Histogram (top) of the yearly number of earthquakes in the Philippines (all Magnitudes) during the period 1964-1991 (International Seismlogical Centre, U.K.). The lower sketch (ING), which visualizes the reference area, shows earthquake epicenters in the 1964-1988 period.*

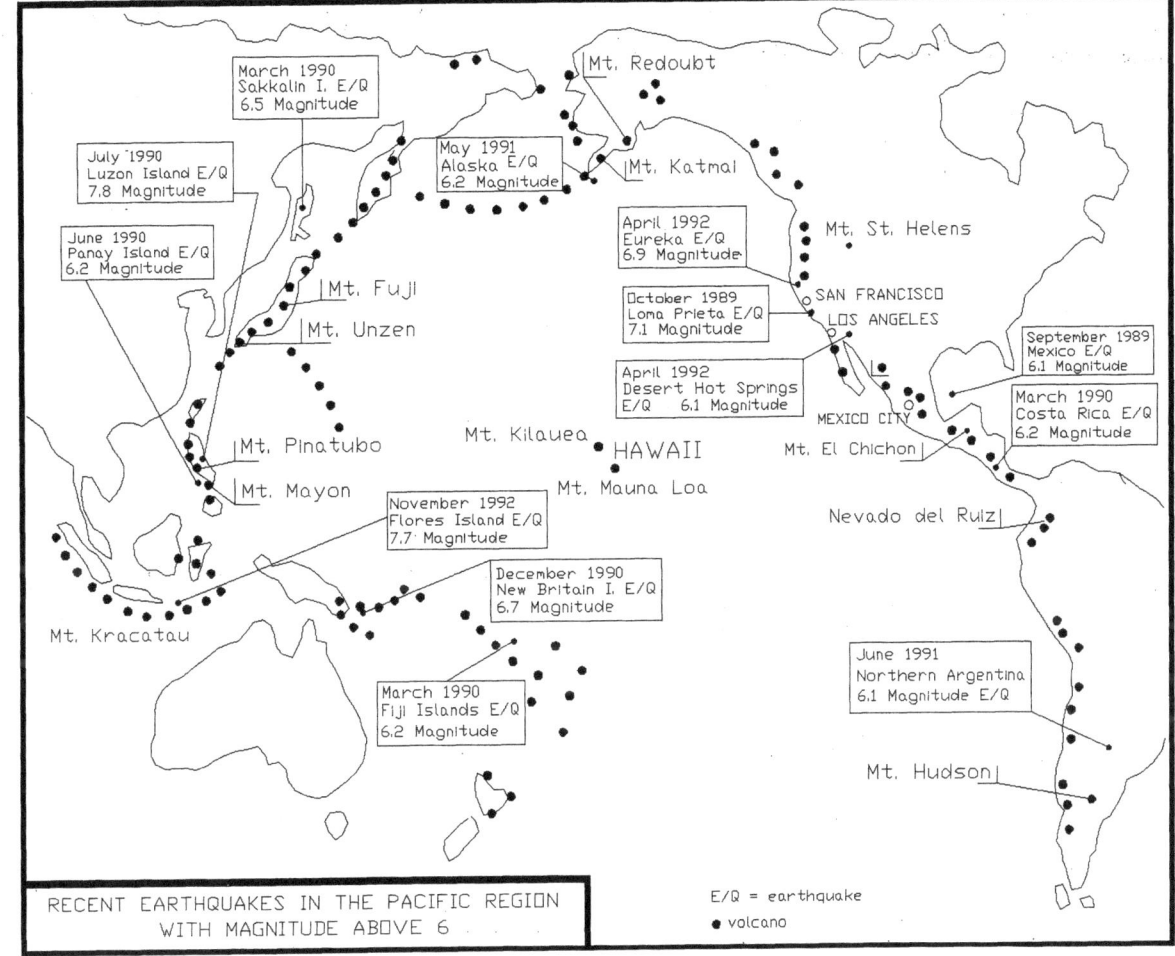

Fig. 4.4 – *Strong earthquakes in the Pacific Plate region, 1989-1992.*

Figure 4.3 (upper sketch) illustrates the number of earthquakes per year in the Philippines (all Magnitudes) during the period 1964-1991. The seismicity of the Archipelago shows a marked increase from 1964 onward, with critical periods (and destructive earthquakes) in 1970, 1976, 1982 and 1990-91 (see Table 2.1).

The location and Magnitude of strong earthquakes in the Pacific during the 1989-1992 period is shown in Figure 4.4. An essential source of information on seismicity in the Philippines is the Catalogue of Philippine Earthquakes, 1589-1899 by William Repetti (1946). This summarizes seismic activity during the period of Spanish occupation. A more recent source is volume IV by SEASEE (1985) which lists all available data up to 1985. The oldest earthquake record concerning the Philippines is dated 1589.

Historical data show that the archipelago was shaken by frequent quakes of various intensities, with many destructive events between 1589 and 1988 (Fig. 4.5). Except for the events during this century, the location of the previous strong quakes is based on damage distribution and other historical information provided by Spanish sources.

Figure 4.6 shows revised locations of epicenters of the major historical earthquakes along the Philippine Fault zone. Because of differences in the interpretation of data and the consequent location of epicenters, Figures 4.5 and 4.6 are not directly comparable.

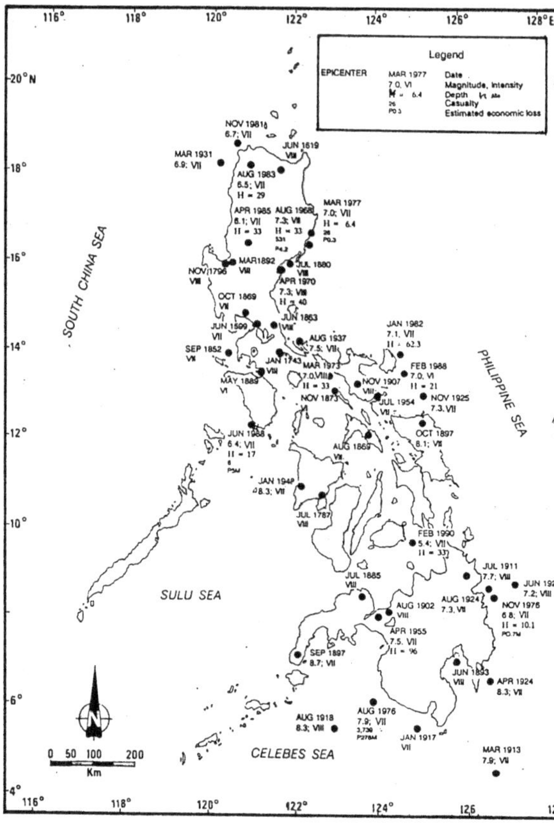

Fig. 4.5 – *Distribution of epicenters of major earthquakes (M › 6 and/or Intensity › VI) in the Philippines, 1599-1988 (Earthquake and Tsunami, PHIVOLCS, 1990). Intensity is based on the Rossi-Forel scale (Appendix C).*

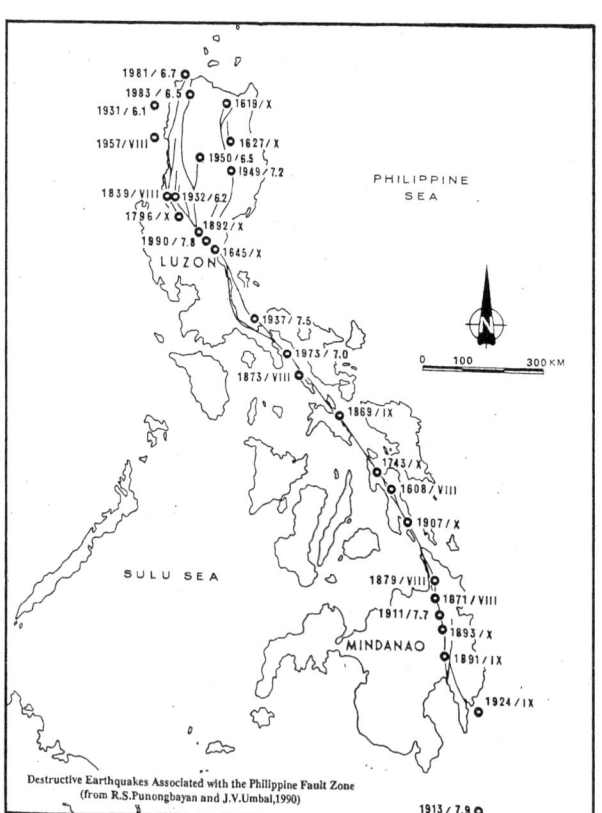

Destructive Earthquakes Associated with the Philippine Fault Zone
(from R.S.Punongbayan and J.V.Umbal,1990)

Fig. 4.6 – *Destructive earthquakes along the Philippine Fault zone during the last four centuries (Punongbayan and Umbal, 1990)*

4.4 Focal mechanisms

A great step forward was made in the clarification of focal mechanisms with the recognition of subduction zones affecting both sides of the archipelago and the strike-slip movement along the Philippine Fault. Additional useful information was derived from the identification of numerous other tectonic lineaments. The revised seismic hazard analysis for the Philippines by Su (1988) includes hazard maps for the Archipelago with different levels of expected horizontal ground acceleration for some given annual probability of exceedence, namely for 0.1% and 0.01% a.p.e. These maps are based on source zones or seismogenic maps which are also incorporated in the aforecited paper.

By combining seismographic data (epicenter, focal depth, magnitude) with the most recent information on tectonics and geology, the Philippine Archipelago was divided into 18 earthquake generators or source zones (Appendix B). The revised approach adopted by Su is based on 3872 earthquakes in the Archipelago from 1964 to 1983 and follows the methodology proposed by Cornell (1968) and Algermissen (1982).

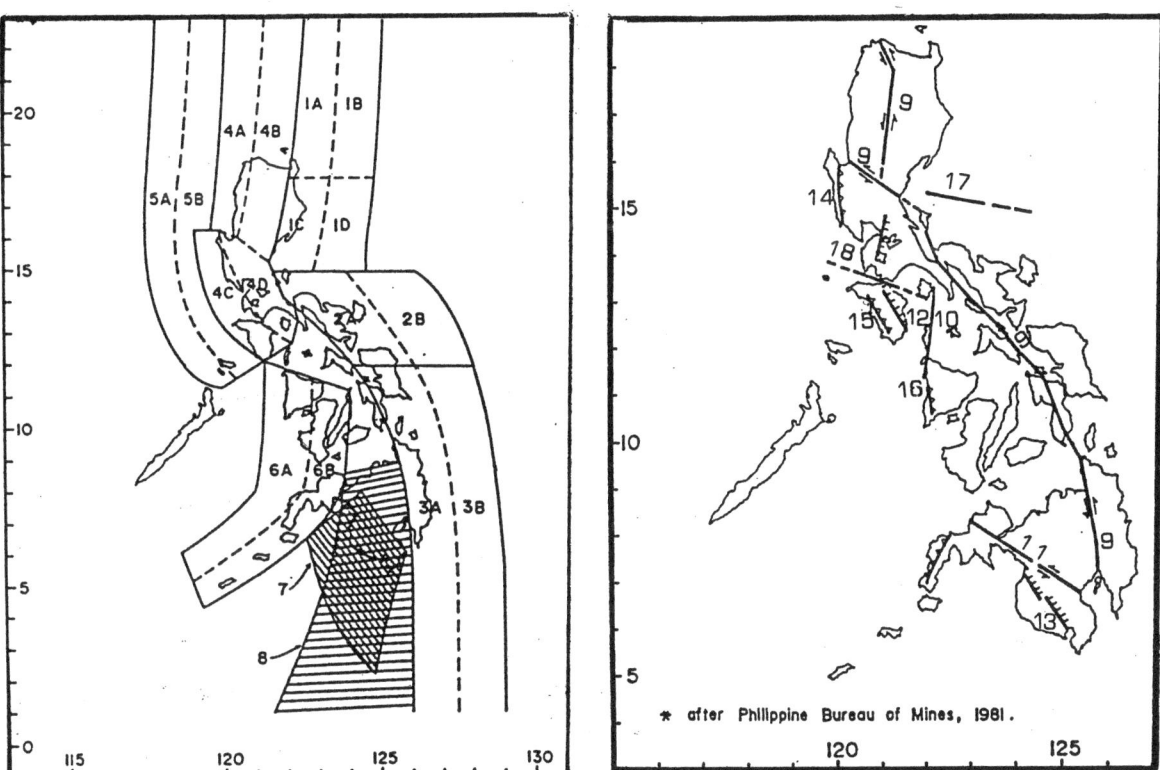

Fig. 4.7 – *Earthquake Source Zones 1 to 8 and 9 to 18 (Su, 1988), Reprinted by permission of Kluwer Academic Publishers.*

Figure 4.7 shows Source Zones from 1 to 8 (left sketch), which are basically related to focal mechanisms of the thrust and strike-slip types, and to mechanisms associated with Benioff zones and trenches. Figure 4.7 illustrates also the major tectonic lineaments in the Philippines treated as finite-width Source Zones from 9 to 18 (right sketch). These include transcurrent, normal and thrust faults (according to the Philippine Bureau of Mines, 1981).

4.5 The 1990 Luzon Earthquake

At 4.26 pm on July 16, 1990, Central Luzon was struck by a catastrophic earthquake having a Magnitude of 7.7 (Intensity VIII, Rossi-Forel scale, Appendix C), with epicenter near Rizal City and a focal depth of 24.8 km (NEIC). The event caused widespread damage and 1, 666 deaths. The worst affected towns were Baguio, Agoo, Dagupan, Aringay and Pura; Tarlac, Cabanatuan, Rizal and Manila were marginally damaged (Fig. 4.1).

The quake, which was felt over an area of about 500,000 sq. km, was accompanied by a 120-km long ground rupture between Gabaldon and Kayapa along the Philippine and Digdig Faults. A sub-surface rupture is believed to have propagated 50-100 km further North, in Northern Luzon (Sharp, USGS-PHIVOLCS Report, 1991), probably along the Abra River Fault (Fig. 4.8), as suggested by the distribution of aftershocks until February 1991.

Two minor surface ruptures, shown in Figure 4.9, were also induced by the quake along the north-western portion of the Philippine Fault and splays in the area between Rizal City and Lingayen Gulf. Also shown in the figure are the areas affected by the destructive quakes of 1645, 1796, 1892 in Central Luzon.

The movement along the major rupture zone, between Gabaldon and Kayapa, was left-lateral with horizontal displacement up to 6.2 m (near Capintalan) and vertical movement between 0.2 and 2.2 m. The horizontal slip mostly exceeded 3 m.

Figure 4.10, by comparison, shows the displacements in California along the San Andreas Fault during the 1979 and 1940 earthquakes. According to Sharp (USGS, Professional Paper 1254) a 5.8 m maximum horizontal displacement, scaled from aerial photographs, was calculated for the 1940 quake in a zone near the Mexican border.

Figure 4.11 shows the alignment of Dalton Pass road

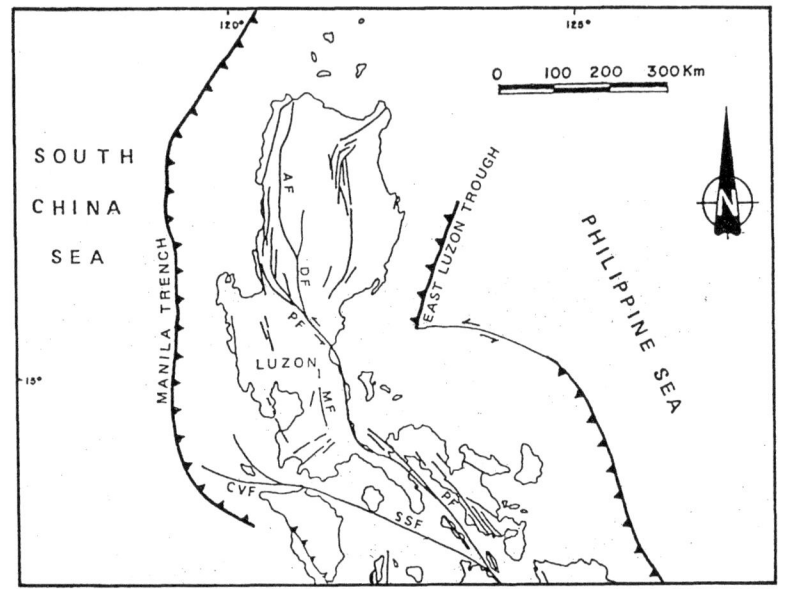

Fig. 4.8 – *Major tectonic lineaments in Luzon (PF = Philippine Fault, DF = Digdig Fault, AF = Abra Fault, MF = Manila Fault), from Punongbayan and Umbal (1990).*

(Pan-Philippine or Maharlika Highway), which runs along the ruptured Digdig Fault segment for about 40 km. Figure 4.12 shows the geometric details of the ground rupture in Figures 4.13 through 4.15, where the fault crosses the road in the section between San Jose' and Santa Fe'.

The ground rupture started SE of Rizal City and proceeded in jerks, following the zones of weakness of the major lineaments. The strike-slip motion occurred along the Philippine Fault segment connecting Gabaldon to Rizal and along the Digdig Fault between Rizal and Kayapa (Fig. 4.9). The bifurcation point, where the ground rupture splits into a northern and a northwestern segment, is located near Rizal. Horizontal and vertical displacements along the minor surface ruptures West of Digdig Fault (in the localities of San Josè-Lupao-San Quintin and San Manuel-Rosario) were minimal (Fig. 4.1), compared with the slip motion between Gabaldon and Kayapa.

According to eyewitness reports in Baguio and Manila, the earthquake consisted of two major shocks with a pause of 2 minutes and 54 seconds (Punongbayan et al., 1991). The first event at 4.26 pm with epicenter near Rizal was classed as having a Magnitude of 7.7 (NEIC) and lasted over 45 seconds. Soon after ground shaking began, recording instruments went off-scale, thus no records are available for the pause and the second shock felt by people in various localities.

Figure 4.16 shows Isoseismal lines, ground rupture and the area affected by landslides (Punongbayan and Torres, 1990).

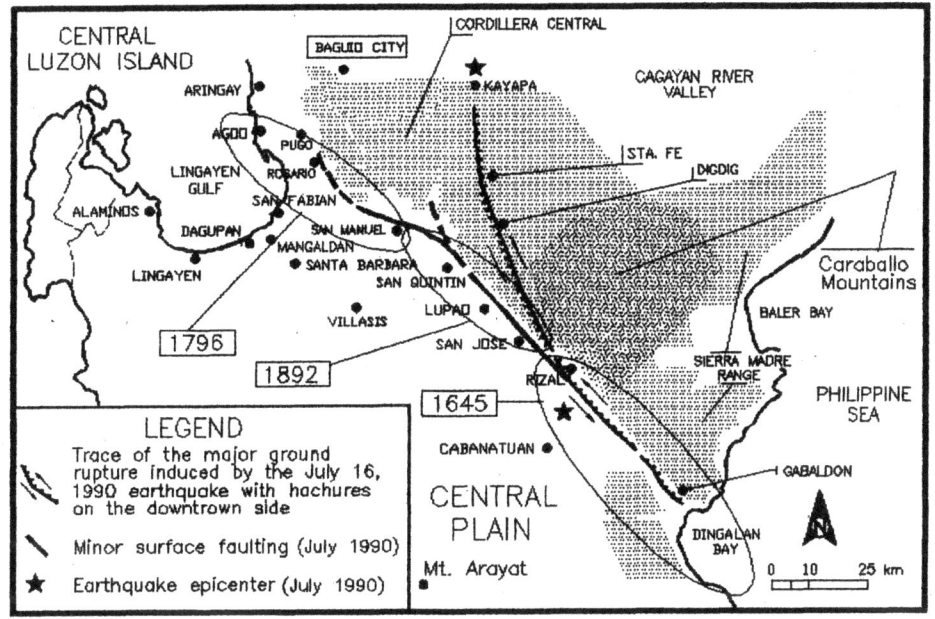

Fig. 4.9 – *Location of the Major Ground Rupture and minor surface faultings in Central Luzon associated with the July 1990 earthquake and the areas affected by the 1645, 1796, 1892 earthquakes.*

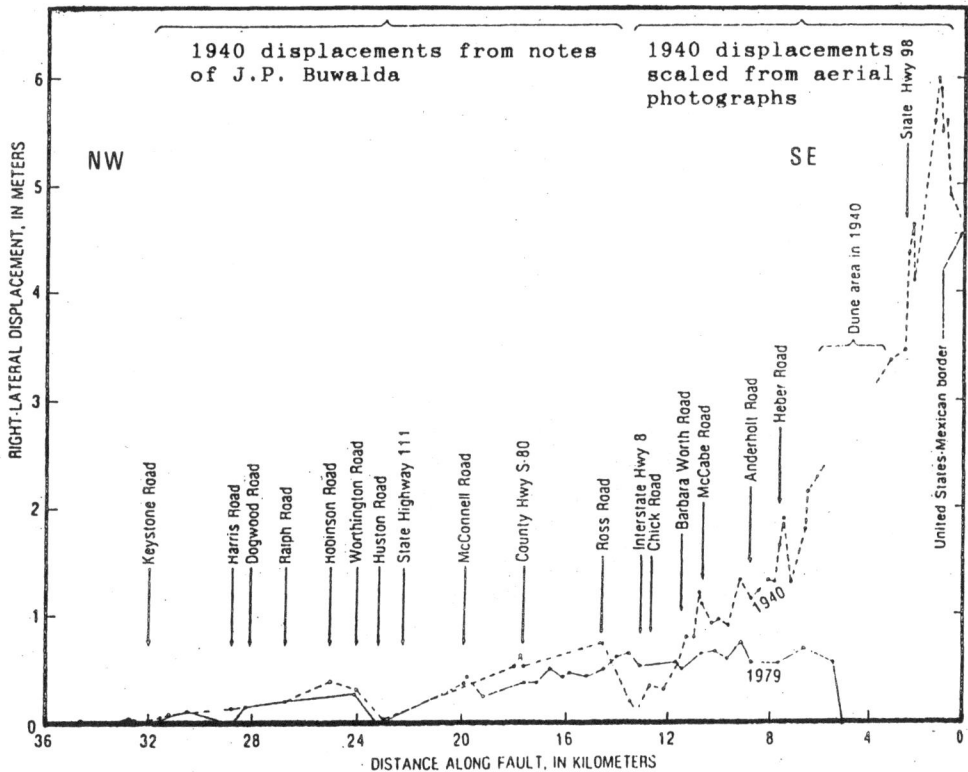

Fig. 4.10 – *Right-lateral displacements in California, along the San Andreas Fault, as a consequence of 1940 and 1979 earthquakes (Sharp, USGS Professional Paper 1254).*

Fig. 4.11 – *Map of Central Luzon with the location of the major roads affected by the quake. The section between San Jose' and Santa Fe', km 165 to km 205, of the Dalton Pass Road (Maharlika Highway) crosses the Major Ground Rupture several times (Figures 4.13 through 4.15).*

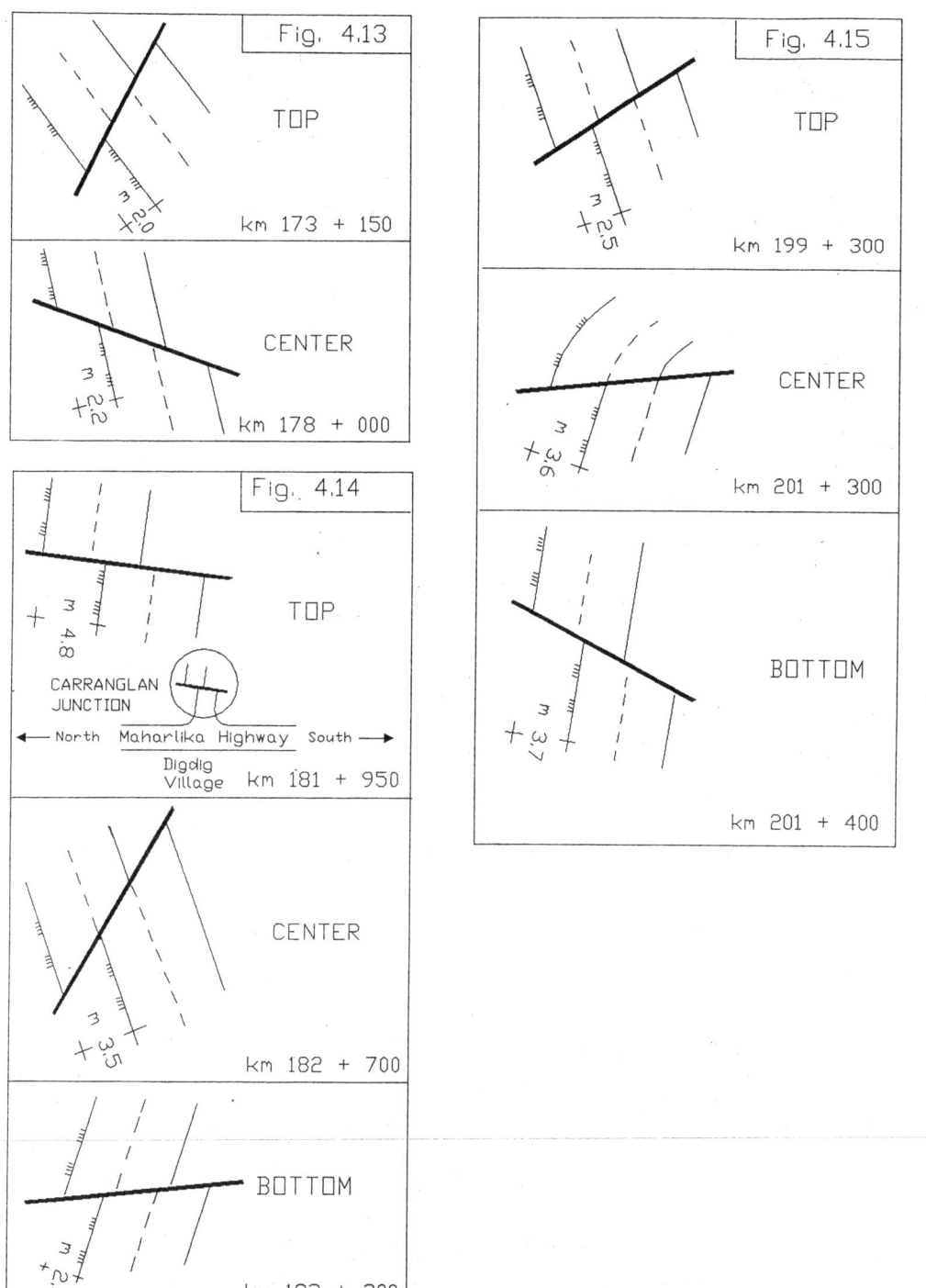

Fig. 4.12 — *Geometric sketches of the broken rigid pavement shown in Figures 4.13 through 4.15, due to the ground rupture (Dalton Pass road). The apparent horizontal slip is related to the margin of the cement concrete slabs.*

Fig. 4.13 – *Ground ruptures damaging the Dalton Pass road at km 173+150 (top), km 178+000 (center) and near km 180 (bottom). Courtesy of DPWH and Katahira.*

Fig. 4.14 – *Faulting in Digdig, at the junction for Carranglan village (km 181+950 of the Maharlika High-way), with 4.8 m horizontal and 1.5 m vertical displacements (top). Fault crossing the road at km 182+700 (center) and at km 183+200 (bottom). Courtesy of DPWH and Katahira.*

Fig. 4.15 – *Fault crossing the Dalton Pass road at km 199+300 (top), at km 201+300 (center) and at km 201+400 (bottom). Courtesy of DPWH and Katahira.*

4.6. The analysis of the earthquake

4.6.1 Data recording and processing

Seismographs in Luzon went off-scale about 45 seconds after the July 16, 1990 ground shaking began, thus earthquake analysis was based on data from outside the Philippines. In particular, seismic signals recorded at three stations in Italy, and made available by the Istituto Nazionale di Geofisica (ING), were used. These Broad Band Stations, under the ING, are part of the Mediterranean Network (MedNet) recently established for monitoring the seismicity in the area (MedNet Workshop, 1990). The system has the advantage of providing an undistorted signal, linear over a very large frequency range. The broad-band signal is digitally sampled from the seismometer and recorded on magnetic tape. The technique was developed in the USA in 1962 (Wielandt, 1991) and the first European seismological station came into operation in 1976 in Germany. The signal of the earthquake can be used for advanced data-processing based on mathematical modelling and the spatial and temporal distribution of the seismic source can be derived along with seismic parameters.

Ground displacement was retrieved first by integrating the broad-band signal (Fig. 4.17) recorded at the Italian stations; then, the characteristics of the ground rupture were determined (Morelli, 1991; Giardini, 1991) through a mathematical procedure based on ray-theory. The earthquake signal was thus decoded and the source-time function and the focal mechanism derived.

A broad comparison was attempted of the initial 230 seconds of ground shaking (time intervals A + B in the lower seismogram of Fig. 4.17) and eyewitness information (Para. 4.5). The first shock that people felt appears to lie within the initial 120 sec shock in the record (time interval A, in Fig. 4.17, bottom). The following pause which reportedly lasted 174 seconds can be partly identified in the seismogram as a clear drop in ground shaking for about 110 seconds (time interval B). The arrival of PP waves at 230 seconds prevents any further attempts at comparison; thus, the second shock with probable epicenter in Kayapa (Fig. 4.1), cannot be identified, due to overlapping of signal and PP waves.

4.6.2 Seismic source analysis

Seismic records from distant stations can thus be used to search for the focal mechanism and source-time function by appropriate mathematical processing. The trace in Figure 4.17 (upper seismogram) shows the Luzon earthquake signal (vertical ground-shaking P wave component) as recorded at L'Aquila Station in Italy. An enlarged detail of the initial 1440 seconds is shown in the lower sketch, with indications of first arrivals of P, PP, S and PKIKP waves. In Figure 4.18 a cross-section of the globe illustrates different travel paths including those related to the four mentioned wave types, while arrival times are shown in the sketch on the right for different angles between the focus, the center of the earth and the receiving station.

In the case of the July 1990 earthquake, a 92 degree angle between the source (Luzon epicenter) and the receiving station (L'Aquila) was used. The cross-section in Figure 4.18 (left) shows the variety of waves (conventionally indicated by capital letters and combinations depending on trajectories) which can arrive at a recording station located at progressively greater distances from the source. The inner core, outer core and mantle play major roles in this respect by reflecting, refracting and diffracting P waves, while the molten outer core entirely blocks S waves. The duration of the seismogram (Fig. 4.17, top) depends on the location of the epicenter with respect to the receiving station and on the type of recording instrumentation: it does not provide any information regarding the shaking time in the epicentral area.

To return to the determination of source parameters, Figure 4.19 illustrates the focal mechanism and source-time function retrieved through the analysis based on ray theory. In detail, the full lines in graphs on the left represent data derived by integrating recorded seismic signals at the three Italian MedNet stations of Villasalto (VSL), Bardonecchia (BNI) and L'Aquila (AQU). Dotted lines show traces derived through the simulation procedure for generating synthetic seismograms. The simulation process is based on four time-domain functions representative of phenomena accompanying seismic wave propagation and on the model assumption of horizontally layered media (Beranzoli et al., 1993).

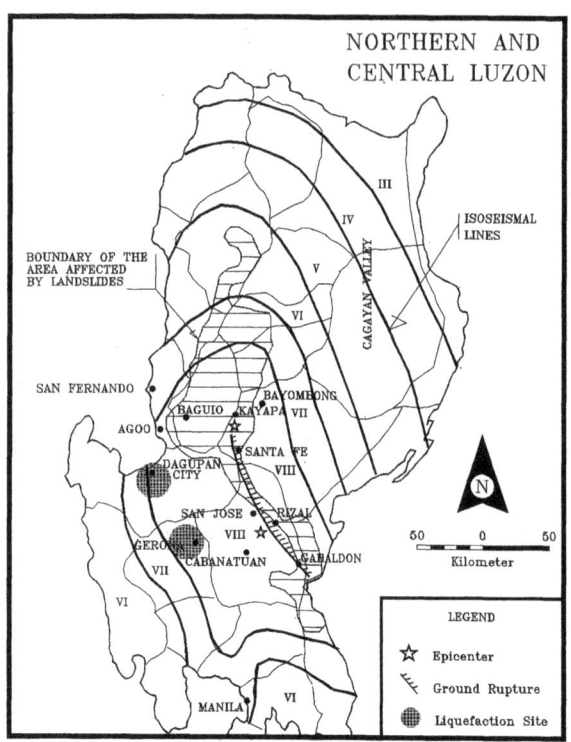

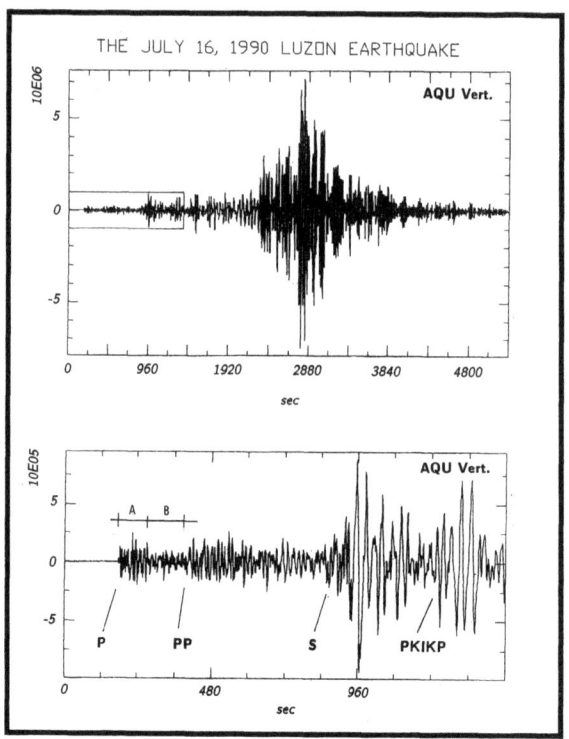

Fig. 4.16 – *Earthquake Intensity Map based on the Rossi-Forel Scale (Appendix C). Adapted from Punongbayan and Torres (1990).*

Fig. 4.17 – *Seismogram of the July 16, 1990 Luzon earthquake (top), recorded at the station of L'Aquila (Central Italy). The lower sketch represents the enlarged portion of the above seismogram for the initial 1440 seconds (Istituto Nazionale di Geofisica, Rome, Italy).*

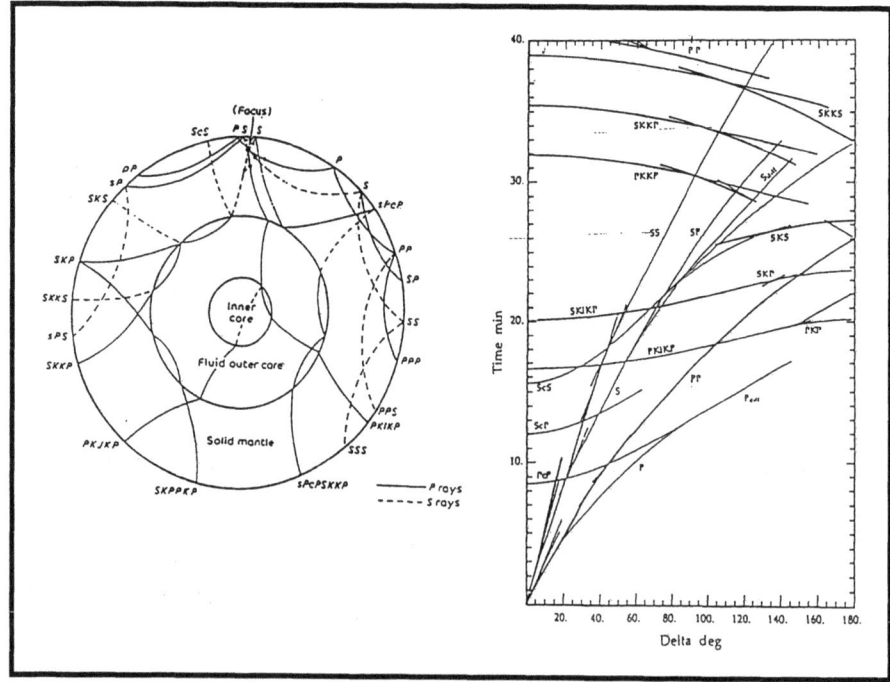

Fig. 4.18 – *The cross-section of the earth illustrates travel paths of different waves originating from the quake focus. Waves are conventionally indicated by capital letters depending on trajectories. The sketch on the right shows the arrival time of these waves as a function of the angle between the focus, the center of the earth and the receiving station.*

In the three 150-second sketches of Figure 4.19, synthetic seismograms (dotted lines) are in good agreement with the traces derived from recorded data. This is interpreted as a successful simulation and, therefore, correct identification of the focal mechanism and source-time function. The upper right-hand sketch in the same Figure shows the focal mechanisms, which is illustrated by the equal area projection of the lower emisphere of the focal sphere. The plane with N147E strike (and SW dip at an angle of 62 degrees) coincides with the direction of the major ground rupture trace along the Philippine Fault near Rizal City (Fig. 4.1), where the first epicenter was located.

The lower-right sketch in Fig. 4.19, which shows the source-time function, indicates that the first major shock was composed of three sub-events for an overall duration of 120 seconds.

Combining recorded data and eyewitness information, the Luzon earthquake was most probably generated by two major shocks, separated by a lower seismicity level which lasted over 120 seconds. The first event (composed of three sub-events) occurred at a depth of 24.8 km; the second epicenter, based on eyewitness reports and damage assessment, can be located near Kayapa, some 30 km E of Baguio.

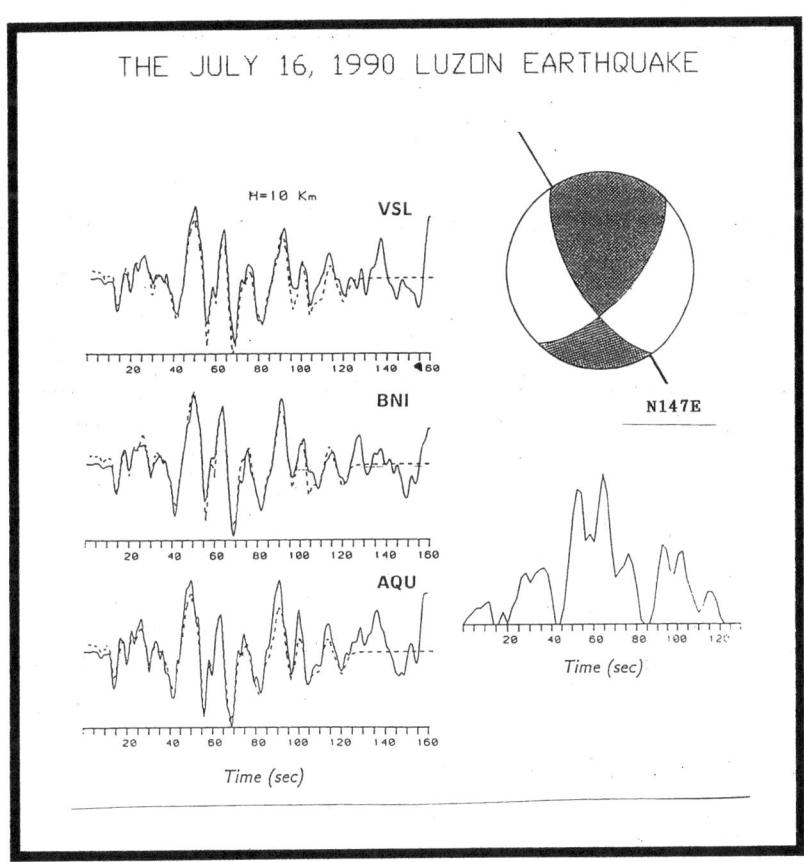

Fig. 4.19 – *Data processing and interpretation, Focal Mechanism Solution and Source Time Function. The sketches on the left show full lines derived from data recorded at the Italian stations of Villasalto, Bardonecchia and L'Aquila, and dotted lines from synthetic seismograms. The agreement between the two is an indication of successful simulation. The equal area projection at top right shows that the identified focal mechanism is of the strike-slip type, and that the predominant direction of motion is lateral. Since the plane with strike N147E (and dip SW 62°) coincides fairly well the direction of the major ground rupture near Rizal (Fig. 4.1) where the epicenter was located, it is interpreted as the fault plane. Dashed and blank zones (ounded by the fault plane and the auxiliary plane) represent compressional and dilatational first motions, respectively, and their position indicates that the motion was mainly left lateral. The lower sketch, the source-time function, shows that the first major shock was composed of three subevents for an overall duration of 120 seconds (Istituto Nazionale di Geofisica, Italy).*

4.7 Damage distribution and subsurface rearrangement

The earthquake caused limited damage in the epicentral area in comparison to Baguio City and the towns along Lingayen Gulf where damage was severe. The destructive effect of the earthquake on the zone west of the ground rupture was exacerbated by the countless landslides which affected the Caraballo Mountains and the Central Cordillera.

In general the damage extended along the ground rupture zone and west of it through the Cordillera. By contrast, the Cagayan River Valley and Sierra Madre Range were marginally involved. According to Sharp (USGS-PHIVOLCS Report, 1990) «the jerky propagation of the rupture and its northwestward trend might have resulted at some degree in an attenuation of the seismic waves in the eastern block and amplification in the western region«. The seismic swarm which followed the quake (Chapter 7) is thought to be related to the compression induced in western Luzon by the 1990 slip. Through a sizable subsurface block readjustment and numerous aftershocks the accumulated stress was released mainly during the period July-October 1990.

One reason for the underground rearrangement was probably that it was difficult for the Cordillera to absorb the compression, due to the presence of deep rooted, massive batholiths with limited and complex displacement potential along the various intersecting lineaments. The block's reassemblage (after the July 1990 quake) was most probably coupled with some uplift of the Cordillera.

4.8 Recurrence period

Based on an average displacement of 2 cm/year along the Philippine Fault, the July 1990 horizontal slip of 6.2 m probably compensated about 310 years of stress accumulation. The last strong earthquake in the same area occurred in 1645, 345 years before the 1990 event. According to Newhall (USGS-PHIVOLCS Report, 1990), however, historical records of the 1645 earthquake mainly deal with Manila, thus the location of this event along the Gabaldon-Rizal segment of the Philippine Fault is uncertain.

The Philippine Fault in Central Luzon splits into a number of splays which can accommodate the crustal shortening along different segments moving at different times. The presence in Central Luzon of other earthquake generators makes the problem even more complex. Some progress in forecasting the probable location of the next strong earthquake in the Central-Northern Luzon region could probably be achieved through further analysis of data and locations of past earthquakes and the movements along active faults.

4.9 Casualties, damage and environmental impact

4.9.1 General

The earthquake intensity map shown in Fig 4.16 gives a general indication of damage distribution in Luzon. Consequences of the 1990 earthquake can be considered under four headings: human losses and injuries; damage to property, infrastructure and services; environmental impact; and the economic damage to the Philippines (described in Chapter 9).

4.9.2 Human losses

There were 1,666 deaths, about 1,000 persons were reported missing and over 3,000 were injured. Most casualties occurred in Baguio City and surroundings. The maximum intensity, VIII, regarded the area between Kayapa, Gabaldon, Tarlac, Dagupan, Agoo and Baguio (Fig. 4.16). A sizable part of the mountainous zone affected by intensities VI and VII is sparsely inhabited so casualties were relatively few. In Metro Manila, where the intensity was VI to VII, people in Makati were terrified by skyscraper oscillations, but no casualties were reported, although numerous old buildings were damaged. Due to the disruption of the road network it took days for the rescue teams to reach some remote areas.

The rainy season, which began soon after the tremor, produced new casualties, some as the result of reactivation of earthquake-induced slides and some because of new slope movements generated by the rains.

4.9.3 Damage to property and infrastructure

Nearly 100,000 houses suffered damage and 40% of them were virtually destroyed. Figure 4.11 is a location map of the area most severely affected by the quake and Figure 4.20 illustrates some striking examples of structural damage in Baguio where a number of hotels and buildings collapsed. Towns in the coastal area south of San Fernando in La Union, along the Lingayen Gulf and in the area between Dagupan and Tarlac (Chapter 5), were damaged by extensive liquefaction. A six storey school collapsed in Cabanatuan, near the epicenter.

Damage to agriculture was mainly limited to the irrigation and drainage system in the area along the ground rupture zone. Due to the vertical displacement along the Fault, the gradients of channels were upset, thus disrupting regular operation of the irrigation network. The supply of crops from the Cagayan River Valley to Manila was subject to delays because of damage to Dalton Pass Road. Additional problems occurred later because of the heavy rains which resulted in the road being threatened by slides for months.

Seven bridges collapsed, eight were seriously damaged and about 20 were affected by various types of lesser damage. Major bridges that were destroyed included the 600-m long truss bridge over the Agno River in Plaridel (Fig. 4.21, bottom) and the five-span Magsaysay concrete bridge in Dagupan City (Figs. 5.2 and 5.10), both of which collapsed due to liquefaction. Many bridge abutments were damaged, approaches suffered marked settlement and piers suffered tilting or settlement.

The road network of Central Luzon (Fig. 4.11) and the Baguio region (JICA Report, 1990) was significantly damaged by the quake. Roads in mountainous areas (Figs. 4.21 top and 6.5) suffered damage due to cut-slope and embankment failures, longitudinal cracks, pavement deformation, differential settlement of fills. Marcos Highway in particular and Kennon Road (along the Agno River), both leading to Baguio City (Fig. 4.11) were largely destroyed by landslides. A segment of Marcos Highway near Baguio, along a very steep hillside, slid down the slope. The situation was made even worse by the onset of the rainy season (Chapter 6). Naguillian and Halsema Roads near Baguio were seriously affected by huge joint-controlled rock slides, mainly through plane failures of sandstone formations. A section of the Maharlika Highway through the Caraballo Mountains (Figs. 4.13 - 4.15) suffered enormous damage.

Roads in the Central Plain, between Dagupan and Tarlac and near Lingayen Gulf, were badly affected by liquefaction-induced effects. The most common features were long longitudinal cracks, waviness of the road bed, separation of shoulders, disruption of the concrete pavement and generalized disintegration of embankments due to the dissipation of the excess pore-water pressure during liquefaction (Chapter 5).

About 60 km north of Manila along the North Super Highway cement concrete girders of a viaduct were displaced about 80 cm with respect to the bridge centerline.

Dams were also damaged in various ways (JICA Report, 1990). Crest settlement occurred at Ambuklao, Binga, Masiway and Pantabangan Dams, the first two about 15 km east of Baguio City, and the second two 10 km east of San Jose' and about the same distance from the first epicenter, near Rizal City (Fig. 4.11). Cracks and landslides were observed in various parts of the basins and along crests.

Ambuklao Dam was silted up to a few meters below water level with an estimated 60 million cubic meters of sediments, a considerable part of which were transported during the two months after the quake. According to the design the dam should have had a useful life of 50 years before complete siltation, but the effects of the quake contributed to shorten this period to a mere 28 years.

River facilities, parapets, and protective structures in general suffered widespread damage due to ground shaking or to lateral spreading induced by liquefaction. Of the public utilities the electricity network was the most badly hit. Numerous poles were tilted and the electricity supply was interrupted for days, and even weeks in some of the remotest areas. Sewerage networks in cities affected by liquefaction were almost entirely disrupted.

Fig. 4.20 — *Badly damaged buildings in Baguio City (top, FRB Hotel and Royal Inn, bottom, Hilltop and Nevada Hotels).*

Fig. 4.21 – Damaged sections of the Dalton Pass Road at km 203+250 (top left) and at km 200+750 (top right), Courtesy of DPWH and Katahira. The collapsed Plaridel Truss Bridge, over the Agno River, in Carmen (Rosales), about 60 km North of Tarlac (bottom).

Fig. 4.22 – *The overtopped Puncan Bridge along the Dalton Pass Road at km 177.*

Fig. 4.23 – *Fan of newly deposited sediments and debris along the Dalton Pass Road.*

4.9.4 Environmental impact

The damage by the quake to the environment consisted mainly of innumerable slides and the destruction of the vegetation cover in the slide areas. Within a narrow strip along the rupture lines the ground surface was visibly broken for over 175 km. Associated effects included landslides, liquefaction, accelerated erosion and further downslope movements of materials loosened by the quake and mobilized again by the rains.

Landslide areas included the mountainous regions of Western and Central Luzon, namely the middle and southern parts of the Central Cordillera and the whole of the Caraballo Mountains. The slides destroyed numerous trees, both where the surface material was mobilized and downslope, where they were covered by debris.

Figure 4.22 illustrates the extent of the damage through the huge quantity of trees which accumulated by August 1990 at Puncan Bridge along the Dalton Pass Road. The numerous slides in mountainous provinces contributed to the uprooting of vegetation on steep slopes where human-induced deforestation had proved impractical. Extensive alluvial fans (Fig. 4.23), produced by accelerated erosion, formed along major rivers and near the periphery of mountains.

CHAPTER 5

SAND LIQUEFACTION IN THE CENTRAL VALLEY INDUCED BY THE 1990 EARTHQUAKE

5.1 The 1990 liquefaction and historical records

A major secondary effect of the July 16, 1990 Luzon earthquake was the liquefaction of sands in a large elongated zone of the Central Plain. The phenomenon had a devastating impact in Pangasinan and Tarlac provinces (Fig. 5.1), where surface soil conditions are conducive to liquefaction.

Cohesionless saturated fine sandy and silty sediments near the ground surface are liable to liquefy under the effect of intense cyclic ground shaking. The pore-water pressure build-up caused in these soils by powerful earthquakes often results in the loss of shear strength and the transformation of the material into a liquidlike mass. The consequences for buildings and other structures in such a case are devastating.

Reconnaissance studies made after the quake showed that liquefaction-related phenomena such as sand boils, fountains, lateral spreading, cracks, fissures and differential settlement had been induced in various areas and isolated spots in the Central Plain west of the major and minor ground ruptures (Fig. 5.1). In addition to the catastrophic damage caused in Dagupan City in the Gulf of Lingayen, liquefaction also affected San Fernando in La Union Province and severely damaged Agoo and Aringay.

The Central Plain in general and the zone around Dagupan in particular are thought to have undergone liquefaction at least twice in historical times. Records are available because the Spaniards built churches all over the country, and soon had to face the tremendous problem of earthquakes. Their solution for minimizing damage was a new architectural style, known as earthquake baroque, consisting basically of buttressed architecture on a massive body structure with squat bell towers. Historical records indicate that two major quakes jolted the zone along Lingayen Gulf, in 1792 and in 1896. In both cases the tremors were felt from Baguio to Manila. Damage was particularly heavy in Dagupan City and environs where eyewitnesses reported sand-boils, fountains and cracks with the emission of water and sand. Major structures, including churches, were badly damaged.

5.2 Liquefaction and related damage in the Central Plain

The 1,500 sq. km strip (Fig. 5.1) affected by liquefaction, the numerous scattered liquefaction-related phenomena in Central Plain and the coastal area between Agoo and Bauang testify to the regional extent of this secondary short-range effect of the July '90 quake. The mid-Luzon tectonic depression, known as the Central Plain (about 30,000 sq. km), was filled with clastic sediments during the Tertiary and Quaternary. The uppermost part of the sequence, consisting of loose to very loose fine sand, reacted to the ground-shaking with liquefaction on a regional scale. Sand boils, lateral spreading, fountains and fissures were reported throughout virtually the entire Plain, even outside the major liquefaction zone. As in the Japanese town of Niigata during the 1964 earthquake

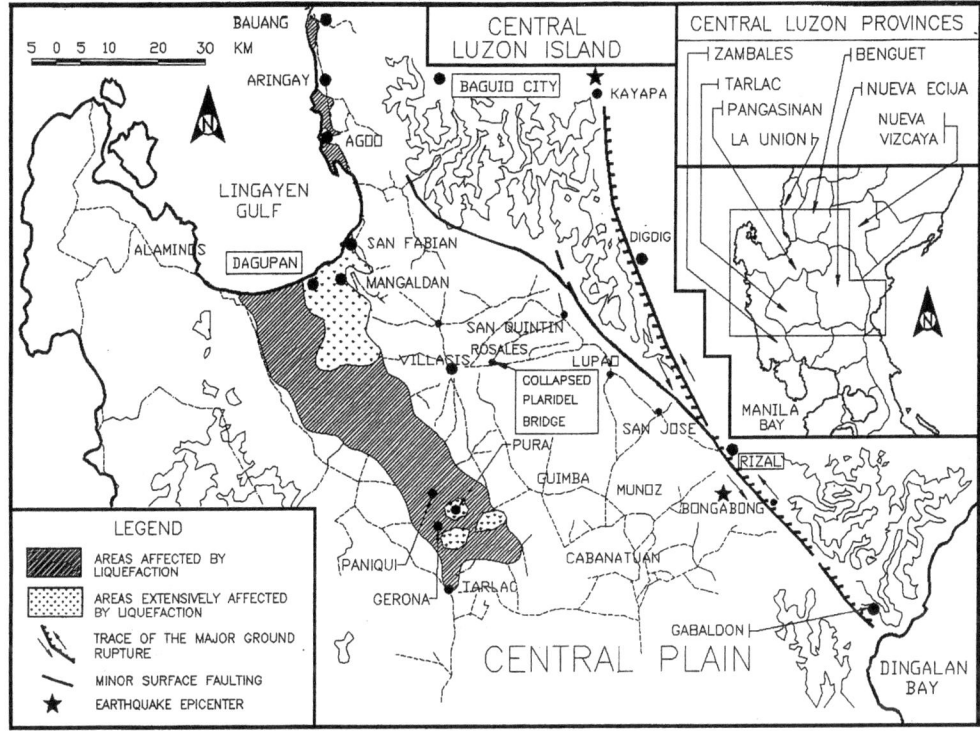

Fig. 5.1 – *Sand liquefaction in the Central Plain and the Agoo-Bauang area (adapted from Punongbayan and Umbal, 1990).*

(Seed et al., 1967), the uppermost sand layer in Dagupan City suffered severe liquefaction during the tremor. Many structures tilted and soon after the quake buildings looked as though they had been floating in a liquid.

According to eyewitness reports, fountains of gas and liquefied sand together with sand boils occurred soon after the beginning of the quake. The gas emissions can be attributed to the presence of decomposed organic matter at various depths within the sand layer, as it was confirmed by borings completed subsequently.

In the countryside near Dagupan City a number of Barangays (small villages) surrounded by water or located on river banks were reported to have subsided severely. Liquefaction in general occurred over a vast zone with an effect comparable to that recorded in Dagupan City.

Extensive liquefaction also explains the severe damage suffered by the road connecting Dagupan City to San Fabian through Mangaldan (Fig. 5.1). The rigid pavement of this road was deformed into an undulating strip several hundreds of meters in length with subsidence of the shoulder on the seaward side (both sides in some cases) ranging from 30 cm to 100 cm. Such behavior, which was also reported in Niigata, is attributable to the fact that the embankment under the rigid pavement acted as a floating shield preventing upward dissipation of the accumulated pore water pressure underneath. As a consequence of this local condition the pore pressure was dissipated along the foot-slopes of the shoulders on both sides of the road, thus resulting in the separation and sinking of the body of the shoulders. For a few months after the quake the cement concrete pavement exhibited marked waviness with a period of about 25 m.

The coastal road directly linking Dagupan City and San Fabian suffered the most severe damage. In some sections it was affected by multiple longitudinal cracks (10-30 m long and 50-90 cm wide) while in other zones the road embankment split into tilted blocks topped by portions of pavement still in place.

Intense liquefaction with sand boils, ground fissuring and subsidence occurred in Agoo, North of Dagupan. Similar phenomena were also reported near the cities of Tarlac, Gerona (collapse of the town hall due to sinking of foundations), Paniqui and Pura (Tarlac Province) where many houses tilted and subsided.

Along the MacArthur Highway, between Gerona and Tarlac, the central longitudinal joint of the concrete pavement separated 2 to 10 cm over a length of several kilometers, with a vertical difference of 5 to 10 cm due to faulting in some limited stretches. About 40 km SE of Dagupan the 650 m long truss of Plaridel Bridge in Carmen (Rosales) over the Agno River (Fig. 5.1) collapsed due to lateral spreading and loss of foundation bearing capacity. The occurrence of sand boils and long fissures was also reported by eyewitnesses along the riverbed which is mainly composed of saturated, loose to medium-dense, fine sediments). Liquefaction was also reported in Manila Port area (Wieczorek et al., 1990) and along the North Expressway leading to Angeles City.

5.2.1 Damage in Dagupan City

The most impressive liquefaction occurred in the commercial area of Dagupan City (Fig. 5.2), which is crossed by two major roads, Perez Blvd. and Fernandez Ave. Many buildings, bridges, road platforms and pavements, as well as electrical and sewerage systems suffered structural damage (top sketch).

In general the surroundings of Perez Blvd. and the zone to the south were hit catastrophically, while Fernandez Ave. was basically affected by severe sinking, the damage being significant in both locations.

The highest density of buildings is concentrated along and between these two roads (Dagupan City proper), the majority of constructions being two to three storeys high with a limited number of four to five storey buildings. In areas with scattered light one-storey concrete or wooden houses liquefaction was not so evident. There was also a significant bulging of the streets and numerous buildings sank and tilted in the most seriously affected part of the city, sand liquefaction being the major cause of foundation failures.

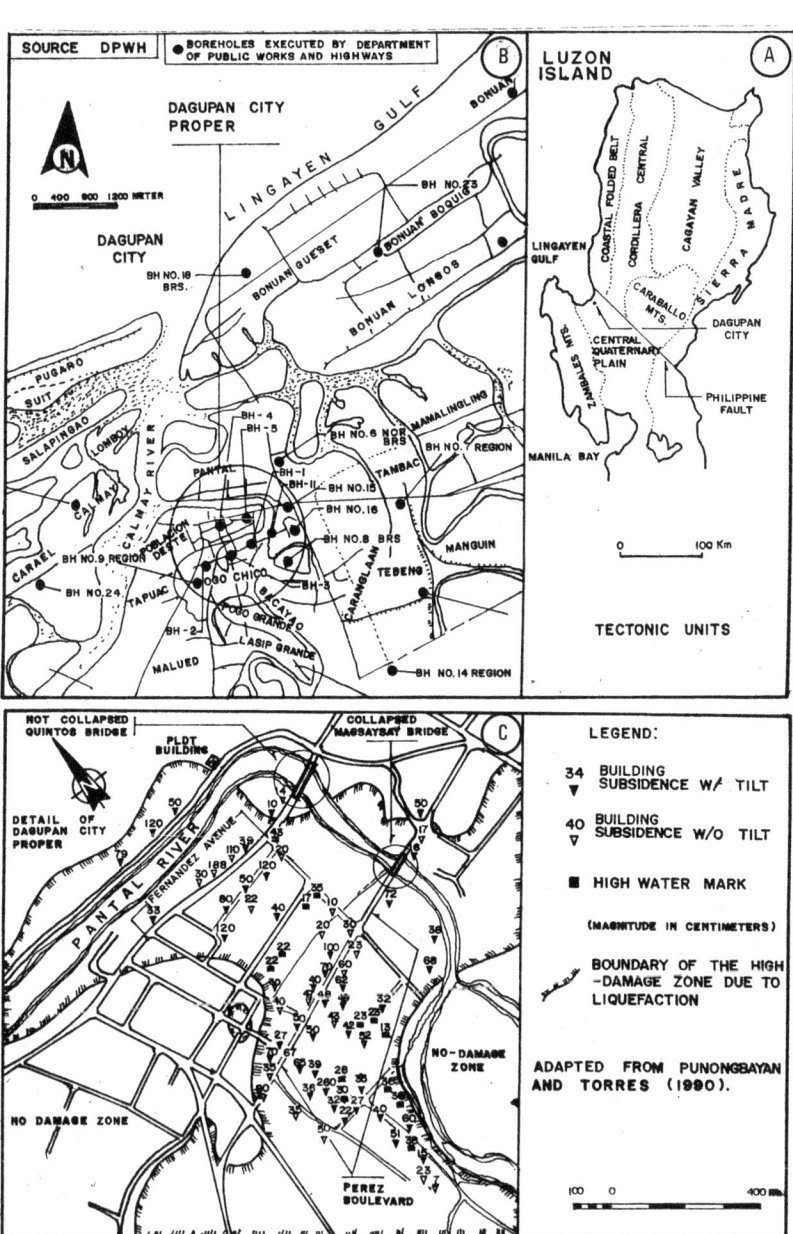

Fig. 5.2 – *Tectonic units in Luzon (A), overview of Dagupan City (B) and details of tilt and subsidence of major buildings in Dagupan City Proper (C). Adapted from Punongbayan and Torres, 1990.*

In Fernandez Ave. contiguous buildings underwent quasi-uniform sinking and moderate tilting for hundreds of meters on both sides of the road (Fig. 5.3, top). Figure 5.3, bottom, illustrates oblique and front views of new, well-kept buildings which suffered severe subsidence (140-180 cm) and tilting in Perez Blvd.

The lack of lateral support from adjacent structures was an essential factor for the severe tilting in Perez Blvd. of two isolated buildings (Fig. 5.4, top) in close sequence (non-uniform sinking 80-150 cm and tilting 12-14 degrees).

The most spectacular case, however, occurred in Perez Blvd. (Fig. 5.4 bottom, left), where tilting was estimated to amount to 15 degrees. The complete rotation and collapse of the structure was prevented by the next-door building. Heavy damage in general occurred in the area between the two major streets in Dagupan City proper, and some houses collapsed in Perez Blvd. Market.

Roads in Dagupan were catastrophically affected by sand liquefaction and the consequent bulging was the major cause of the destruction of the rigid pavement. The swelling of city streets was due to the light weight of the road structure compared with the heavier nature of the buildings. Figure 5.4, bottom right, shows a small truck which has sunk into the roadside due to road bulging and associated subsidence of the shoulders. Concrete pavements of houses and courtyards were disrupted almost everywhere.

Figure 5.5 shows the damaged pavement of the gasoline station on Perez Blvd. where the underground fuel tanks floated up.

Figure 5.6 illustrates Magsaysay Bridge on Perez Blvd., which collapsed during the quake. Four of the seven spans slid into the river, while the remaining three were significantly affected by the quake due to the tilting of the supporting piers. Lateral displacement of soils towards the center of the channel due to liquefaction was responsible for this tilting from the right and left banks towards the middle of the river. Figure 5.6 (bottom) shows the horizontal slip of concrete beams at the same site and the tilted sheet-piled cofferdam built before the quake with a view to widening the bridge.

In contrast, Quintos Bridge about 350 m downstream on Fernandez Ave. (Fig. 5.2) and structurally similar to Magsaysay Bridge suffered no damage. Besides the failure of building foundations and the bulging of roads, electricity poles were tilted and part of the power network in the city area was disrupted. Road bulging also affected the sewerage system.

Statistics for the City (Tokimatsu, Midorikawa and Tamura, 1991) as of July 31, 1990, indicate that 200 commercial and residential buildings were totally damaged and over 400 buildings partly damaged. More than 1,200 houses were destroyed beyond repair and about 6,000 houses partially collapsed. Figure 5.7 shows the damage statistics for about 120 reinforced concrete buildings in the area affected by the quake compared with the ones for Niigata as a result of the 1964 earthquake.

Over 90% of buildings were sampled between Fernandez Ave. and Perez Blvd. Only a few had piled foundations and about half of them tilted more than 1 degree in the liquefied zone. Figure 5.7 shows that both Dagupan City and Niigata were affected by similar damage. Differential settlement and lateral displacement of foundations in both cases were the major causes of failure.

Repeated visits to Dagupan City between 4 and 8 months after the earthquake showed that some buildings which first tilted during the July 1990 tremor continued to tilt for months finally being declared unsafe. This means that, under the unbalanced weight of already inclined structures, continuous uneven sinking persisted for a considerable period. A number of heavy structures affected by sinking and tilting were responsible for damage to nearby buildings. In some cases they leaned against them and in others they influenced the behavior of their foundations.

Adjoining buildings on both sides of the road were responsible for the bulging of streets and consequent damage to pavement and buried utilities. After the quake large areas in Dagupan City and the surrounding countryside were inundated as a result of the dissipation of pore water pressure.

5.2.2 Damage in Pura City

The city of Pura in Tarlac Province, which is located east of the MacArthur Road about 4 km from Paniqui (Fig. 5.1), was also affected by the severe liquefaction of surficial sands. The damage to structures was not as spectacular as in Dagupan City, due to the smaller size of the city and the ab-

Fig. 5.3 – *Quasi-uniform sinking (150 cm) of contiguous buildings (top left) and uneven sinking and tilting (top right) on Fernandez Ave. Sunk and tilted buildings on Perez Blvd. (bottom), oblique and front views, with a differential settlement of 140-180 cm.*

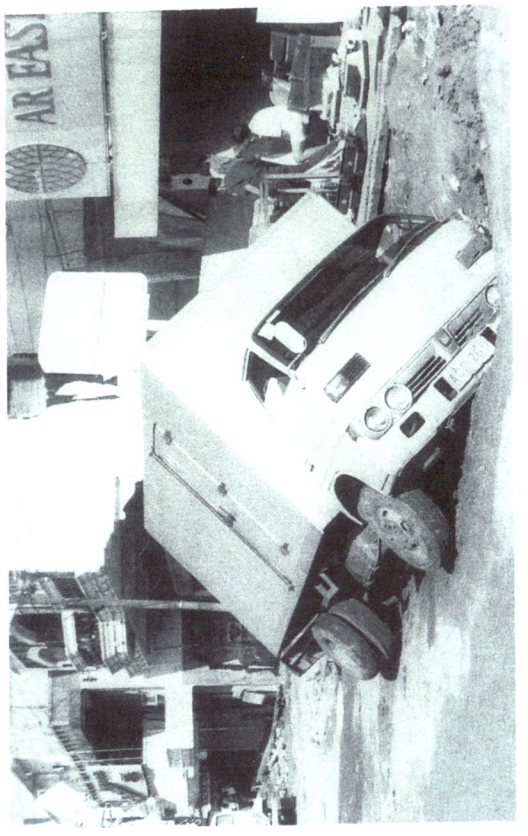

Fig. 5.4 – Isolated consecutive buildings on Perez Blvd. which suffered 150 cm differential settlement towards the roadside (top). The tilting continued for months after the quake. Isolated tall building in Perez Blvd. (bottom left) with 15 degrees tilt due to severe foundation failure. Further rotation and collapse of the structure was prevented by the house next door. Truck sunk into the roadside (bottom right) due to road bulging and subsidence of the shoulder because of liquefaction (Courtesy of R. Punongbayan).

Fig. 5.5 – *View of the gasoline Station on Perez Blvd. (near the collapsed Magsaysay Bridge) where the cement concrete pavement was entirely disrupted by liquefaction and, locally, by the buried tanks floating up. Details of the upheaval of the pavement (center and bottom) due to the floating of tanks (Courtesy of R. Punongbayan).*

Fig. 5.6 – *Collapsed Magsaysay Bridge seen from the left-hand river bank (top). Sunk and 12-degree tilted sheet-piled cofferdam (bottom). The horizontal displacement of concrete beams, due to lateral spreading and consequent tilting of the pier towards the center of the riverbed, exceeds 1 meter.*

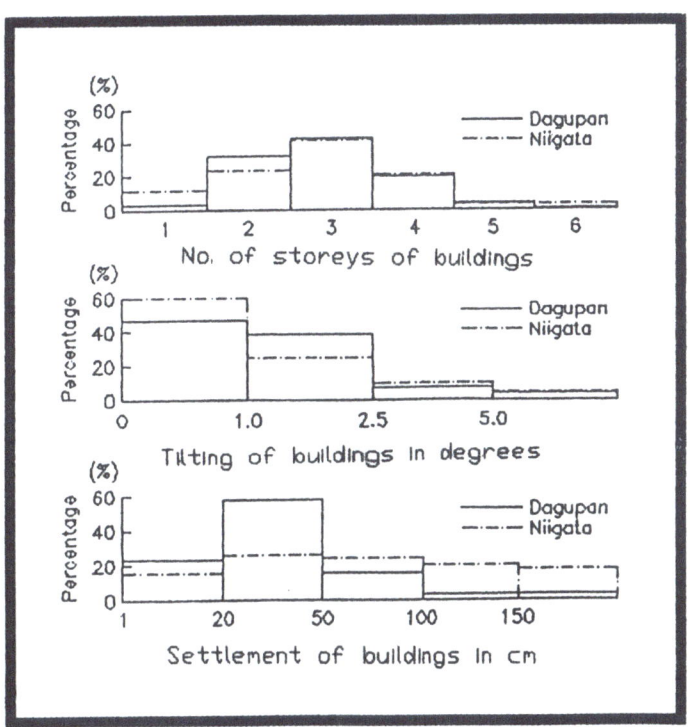

Fig. 5.7 – *Comparison of damage in Dagupan City during the 1990 quake with that in Niigata (Japan) as a result of the 1964 earthquake (Tokimatsu et al., 1991).*

Fig. 5.8 — Tilted houses in Pura City Proper (top). Isolated house outside Pura City with over 1.3 m of uneven settlement, and tilting and the bulging of the pavement inside the house (bottom).

sence of adjoining buildings over two storeys high. Eyewitnesses reported sandboils and fissures throughout the urban area.

The liquefaction of soils hinders the propagation of highly destructive shear waves. Thus, the damage to buildings and structures in general was mainly due to tilting and huge differential settlement, there being only a few signs of cracks, the concrete framework often remaining intact. Figure 5.8 shows some typical examples of tilting and differential settlement associated with the July 1990 earthquake. The road network was entirely disrupted, with long embankment undulations, longitudinal cracks and separation of shoulders.

5.3 Evaluation of the liquefaction potential in Dagupan City

Most of Dagupan City and the surrounding countryside facing the Gulf of Lingayen is a flat lagoon plain filled by loose sediments during the Quaternary (Fig. 5.1). Dagupan City, in particular, was built on a delta with some minor expansion zones created by land reclamation and construction of artificial cut-offs. Abandoned meanders of the Pantal River provide a useful indication of stream migration (Punongbayan et al., 1990; Torres et al., 1990) during the Quaternary (Fig. 5.9). The location of the old meanders coincides with the high-damage zone where severe liquefaction occurred. Loose deltaic deposits characterize the city area, the top sand layer (12-18 m) overlying more than 40 m of clay with silty sand intercalations. The uppermost 4 m of sand may be locally replaced by clay including earthfill material. The water table is usually located at a depth of 1 to 1.5 m.

Figure 5.2 shows: sketch A) a map of the major tectonic units in Luzon, sketch B) the liquefied area in Dagupan City and the two most affected roads in the city proper, namely Perez Blvd. and Fernandez Avenue, sketch C) the detail of the area severely affected by liquefaction with data on building subsidence and tilt, and the location of the major city bridges (circled) on the Pantal River. The

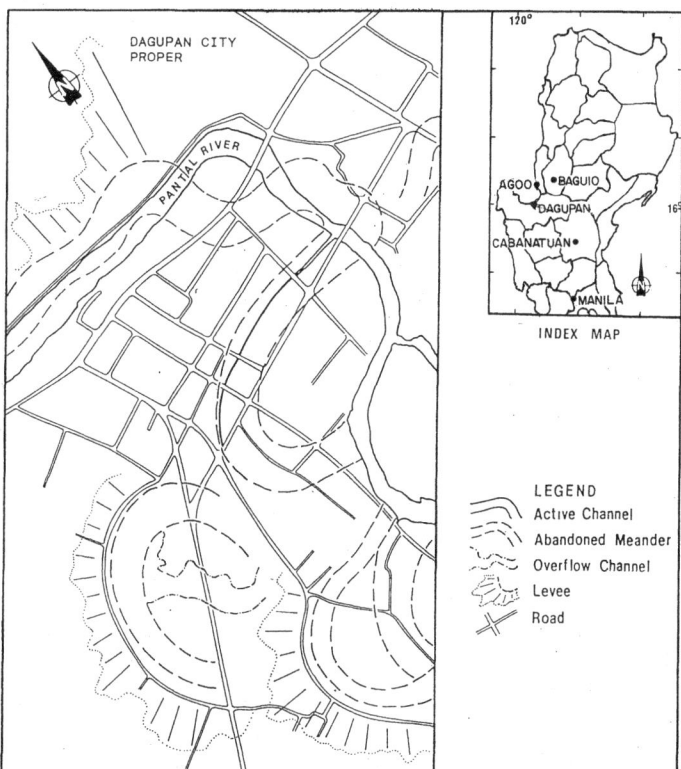

Fig. 5.9 – *Abandoned channels of Pantal River in Dagupan City Proper. The meandering of the river, during its Quaternary migrations, and the related deposition of fine sediments coincide quite well with the location of areas severely affected by liquefaction and high structural damage (Punongbayan and Torres, 1990).*

location of the PLDT building, which did not suffer any damage, is indicated northwest of the undamaged Quintos Bridge (sketch C).

Soil properties of the areas of the collapsed Magsaysay Bridge and the undamaged PLDT building were used for the calculation of the liquefaction potential. Figures 5.10 and 5.11 show the soil profiles, the damage zones and other relevant data along Perez Blvd. and Fernandez Ave.

In Perez Blvd., where the heaviest damage occurred, and in Fernandez Ave. which also heavily suffered the consequences of liquefaction, a number of boreholes and SPT were executed after the quake and laboratory tests were run on materials from various depths. Borehole data and laboratory test results relating to investigations conducted before the quake in the PLDT building area were also made available.

The simplified procedure proposed by Tokimatsu and Yoshimi (presented at the 1984 Eighth World Conference on Earthquake Engineering in San Francisco) was adopted for evaluating the liquefaction potential. This method basically follows the Seed and Idriss (1982) procedure, but has the advantage of introducing the effect of fines. Many investigators recognized the fact that soil vulnerability to earthquake-induced liquefaction is strongly influenced by the presence, quantity and type of fines (silt and clay). Based on field behavior and laboratory testing on undisturbed sand samples obtained by the freezing method, Tokimatsu and Yoshimi proposed an equation to compute the factor of safety F_l (liquefaction resistance factor). This is the ratio between the resistance of soil elements to dynamic loads and the dynamic loads induced by the earthquake motion. When F_l equals 1 or less the soil is likely to liquefy, above 1 liquefaction does not occur.

For a given earthquake the procedure is usually applied to close-by sites with extreme liquefaction and no-liquefaction conditions, high structural damage and no-damage at all, respectively. The

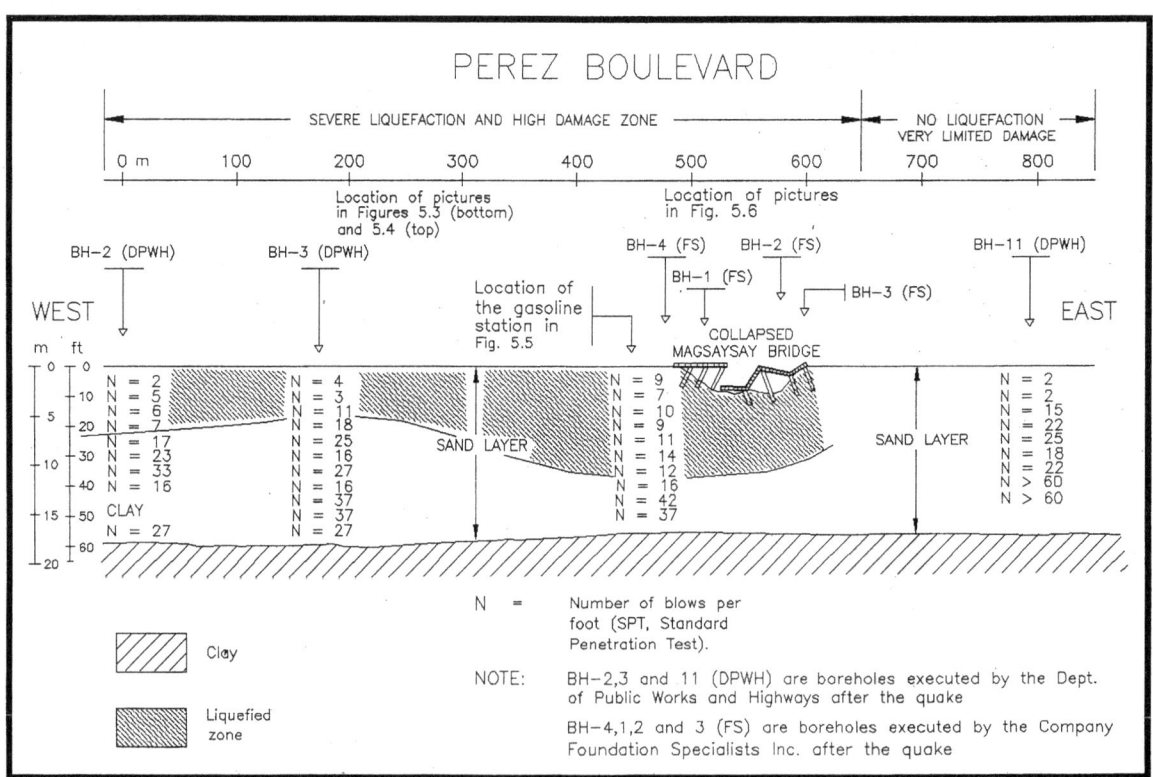

Fig. 5.10 – *Cross-section along Perez Blvd. where the major damage and the most impressive liquefaction effects occurred. Shown in the section are SPT (Standard Penetration Test) data and the layer which underwent liquefaction during the quake.*

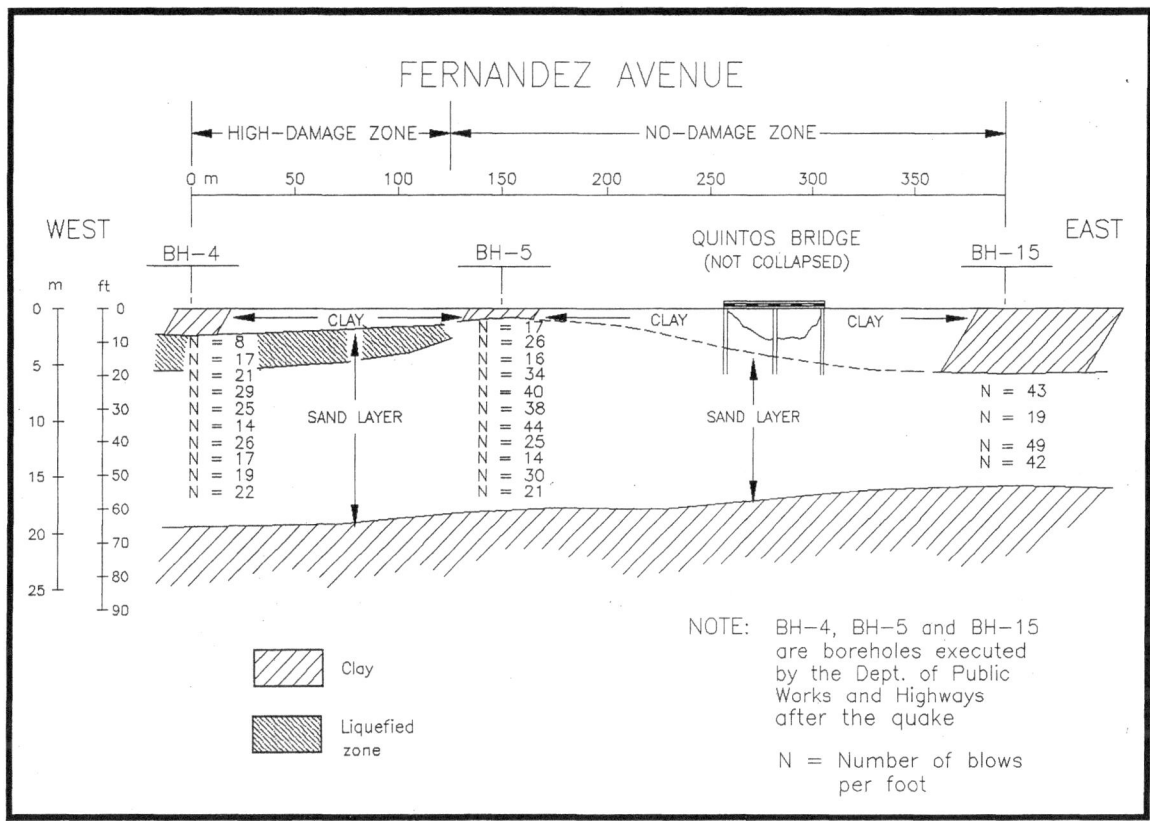

Fig. 5.11 – *Cross-section along Fernandez Avenue. Roadside buildings along the high damage zone were affected by quasi-uniform sinking. The section shows SPT data and the layer affected by the liquefaction.*

maximum ground acceleration, which satisfies in the Tokimatsu and Yoshimi equation both these extremes, is considered the value actually mobilized by the earthquake.

The outlined procedure was applied to the collapsed Magsaysay Bridge (Figs. 5.3C and 5.10) and to the undamaged PLDT building which is located near Quintos Bridge (Figs 5.3C and 5.11), both outside the liquefied area. The two bridges on the Pantal river and the building rested on piled foundations.

In terms of safety factors it was found that F_l was less than 1 for the top layer at the Magsaysay Bridge site and mainly greater than 1 for the same stratum at the PLDT Building.

In conclusion, the entire layer from ground surface down to a depth of 7 to 10 meters underwent liquefaction along a 600 m long cross-section in Perez Blvd. during the quake (Fig. 5.10). Severe damage involved most buildings in this area, several of them being affected by considerable uneven settlement and spectacular tilting. It is worthwhile to recall that the collapsed Magsaysay Bridge is located very near the gasoline Station where buried tanks floated up and the structural damage was most severe.

A thin liquefaction zone was identified along Fernandez Ave. (Fig. 5.11), where buildings mainly underwent a quasi-uniform settlement (90-120 cm).

In both commercial zones of Perez Blvd. and Fernandez Ave., and the area between them, however, 90% of the buildings were condemned, regardless the extent of subsidence due to liquefaction.

CHAPTER 6

LANDSLIDES ASSOCIATED WITH THE LUZON EARTQUAKE

6.1 Introduction

The July 16, 1990 earthquake not only caused surface faulting and liquefaction on a regional scale, but also unprecedented slope failure in central and northwestern Luzon with innumerable landslides in the Cordillera Central and Caraballo Mountains (Fig. 6.1). Out of the 1,666 earthquake-induced casualties nearly 450 were due to landslides and an additional 45 to new slope failures and streamflows activated by monsoon rains during the period August-October 1990.

As a secondary short-range effect of the earthquake, numerous shallow-seated slides were triggered in the mountainous provinces of Luzon. In addition to the catastrophic damage and the environmental impact caused by the countless scars, the quake generated an enormous quantity of loosened sediments with further potential for mobilization by monsoon rains. Part of this disaggregated mass, still lying near footslopes, will be a threat to people and a hazard to property, infrastructure and the environment for years to come. The regional reshaping of landforms induced by the quake and by the subsequent rains is one of the most striking examples of rapid geomorphic evolution of this century.

Based on a number of site visits along Dalton Pass Road and in the Baguio region, a preliminary estimate of landslides associated with the quake puts the number at nearly one hundred thousand within the affected area. A detailed inventory of slope instability phenomena is far from complete, due to the vastness of the territory involved and the inaccessibility of some areas. A comparative study of three sets of satellite images taken before and immediately after the quake and at the end of 1990 could help in the identification and mapping of seismically-induced slides and slope failures triggered by the monsoon rains.

Slope destabilization by ground-shaking continued throughout the August-October 1990 rainy season with peaks during the major typhoons. The mass mobilization activated by the quake clogged rivers, created small dams which later washed out, raised

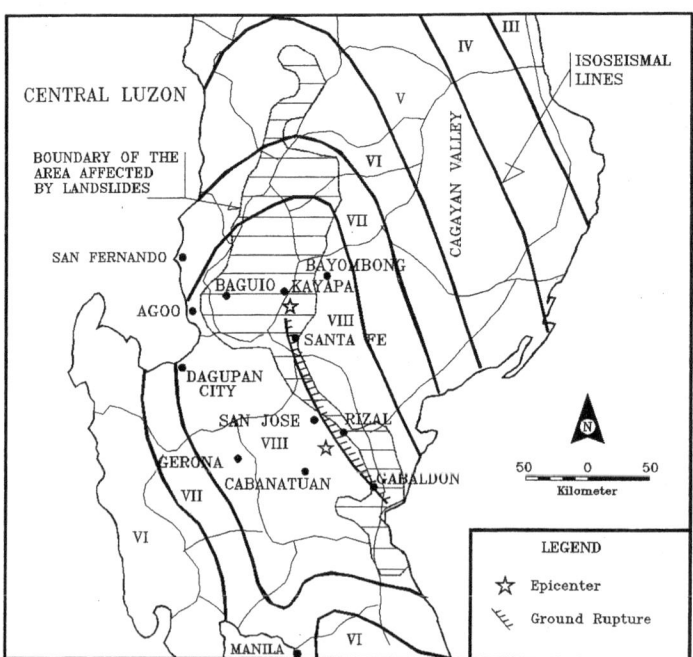

Fig. 6.1 – *Isoseismal Map of Luzon 1990 earthquake and location of the area affected by landslides. Intensities are based on the Rossi-Forel Scale (Punongbayan and Torres, 1990).*

river beds, shaped banks and destroyed bridges and minor structures, and finally built up extensive alluvial fans in the flatlands fringing the mountains.

Slides affected roughly 10,000 square kilometers and, due to the variety of local topographic and geological conditions, included nearly all known types of slope failure. Figure 4.11 shows most of the area severely affected by slides (Benguet province, Caraballo Mountains plus the zone along the major ground rupture) and the location of the main roads in Central Luzon.

The area affected by slides (Fig. 6.1) has been estimated on the basis of PHIVOLCS data and information from various other sources, as well as the author's visits to the Caraballo Mountains (along Dalton Pass Road) and Baguio Province (along Kennon and Halsema Roads, Marcos Highway and the road to Ambuklao Dam).

6.2 Tectonic environment

The Cordillera, which is dissected by the splays of the Philippine Fault, is divided into some sub-ranges and interposed valleys, with an approximate, but visible, N-S pattern. The drainage network basically follows the tectonic lineaments as these evolved during Miocene through Quaternary. Thus, the range is characterized by incised valleys and steep sided crests with some undulating plateaus. The core of the chain consists of large granitic and granodioritic batholiths outcropping at various locations.

The slope destabilization process in the Cordillera Central started by the quake was greatly facilitated by the tectonic history of the range. During the emplacement stage, intruding batholiths metamorphosed, intensely fractured, folded and uplifted surrounding earlier formations (shallow marine deposits, mainly limestones, conglomerates, sandstones and shales), partly destroying their original structure.

Volcanic activity during and after the orogenesis further complicated stratigraphic sequences with the overlapping or lateral formation of pyroclastics, basaltic flows, andesitic and dacitic lavas. Rock formations at the periphery of batholiths, already affected by lateral variations and disturbed by tectonics, became highly unstable and prone to accelerated degradation.

Batholith upheaval episodes from Miocene through the Quaternary generally resulted in a marked decline in the mechanical properties of surrounding rocks. Intrusive bodies (granites and granodiorites), which are generally least affected, during the emplacement process were also faulted and fractured with the formation of shear zones and joints. The presence of discontinuities and shear zones heavily contributed to slope destabilization during the quake.

At the present time the evolutionary process of the Cordillera is greatly affected by the deformation induced by the double-sided subduction in Luzon. Part of the subduction-related crustal shortening is converted into strike slip motion along the Philippine Fault; part of it is turned into the uplift of the range, with the consequent rejuvenation of its drainage system, progressive steepening of slopes, accelerated erosion and undercutting at the toe of scarps by the numerous streams. Uplift and deforestation in the mountains of western Luzon markedly contributed to landslide hazards during the earthquake.

The Caraballo Mountains, which are located between the southern terminus of the Cordillera and the Sierra Madre Range, are generally characterized by lower relief contrast and hilly landforms. The range consists of andesitic rocks with intercalations of sandstones and shales, basaltic lavas and dolerites resting on a basement of pre-Tertiary tonalites and schists. Countless shallow-seated landslides occurred there during the quake, particularly in the area along the major ground rupture which also marks the courses of the Talavera and Digdig rivers.

6.3 Landslide distribution, volumes and geometry

Most of the slope failures were concentrated in the mountainous zones south, west and along the major ground rupture and its probable underground extension, namely part of the Central Cordillera,

Caraballo Mountains and a small zone of Sierra Madre, near Dingalan Bay. The northern third of Cordillera and the area east of the rupture line (which includes the eastern Cordillera, the Cagayan River Valley and the northern Sierra Madre Range) were marginally affected. The lower quake Intensity, the smaller contrasts in elevation and the presence of a nearly intact rain forest contributed in particular to limiting damage to the slopes of the Sierra Madre.

The concentration of slides along and west of the ruptured fault segment was influenced by the following factors:

a) the uplift history of the Cordillera batholiths, considered responsible for marked geomorphologic contrasts and, hence, for the presence of numerous steep slopes locally topped by shattered rocks.

b) the rigid behavior of the rock basement of Cordillera, made of deep-rooted batholiths;

c) the presence of an intricate system of faults and sub-faults in the most affected zone; relative motion probably occured during the quake along many of these tectonic lineaments;

d) the widespread deforestation in the Cordillera and Caraballo Ranges.

The Caraballo Mountain slopes, in general, behaved the same way as part of the Cordillera, with numerous shallow slides. This can be attributed to the presence of uniformly weathered zones in both ranges, associated with low topographic contrast. Figure 6.1 shows isoseismal lines based on the Rossi-Forel scale (Appendix C), landslide distribution and the location of major ground rupture and epicenters. Slides predominantly occurred in the Intensity VIII and VII zones, though there was a marginal involvement of the zone VI which appears to mark the damage threshold. Zone VIII includes major and minor ground ruptures, the epicenter near Rizal and the probable second epicenter near Kayapa.

The distribution of slope movements in the Cordillera was not homogeneous, being strongly influenced by topographic variations and the local condition of the rock formations both in terms of geology and tectonics. In some areas of this range nearly 60% of the slopes collapsed. In the Caraballo Mountains almost no slope withstood the tremors.

About 100,000 landslides were probably induced by the quake, including those newly triggered and a huge number of old reactivated slides. Largely predominant were shallow-seated slope movements, ranging from small to medium size, with volumes mostly in the 1,000 to 5,000 cubic meter range. Due to the extreme variations in geologic and topographic conditions, landslides were unevenly distributed. Pictures of the strong earthquakes in Guatemala (1976) and New Zealand (1968), from Harp et al., 1981 and, Crozier, 1986, respectively, show a generalized landscape instability similar to that of July 1990 in Luzon.

A very broad estimate has been made of the volume of material mobilized by the quake and by the following rains (second half of 1990). Through a number of site visits to Baguio province and surroundings, to Dalton Pass and the Gabaldon area, about one third of the entire zone affected by landslides was inspected, numerous pictures were taken and available information on inaccessible zones was collected.

Based on the rough figures of 100,000 slides and 4,000 cubic meters per slide (average of small, medium and large ones), about 0.4 cubic kilometers of landscape materials are likely to have moved during the quake. This large volume, which provides some idea as to the vast scale of landscape denudation, was partly mobilized again during the July-September 1990 rains.

Though a number of large deep-seated slides were observed, the great majority consisted of surficial mass movements having a triangular elongated shape and a thickness ranging from 0.5 m to 2 m. In most cases these very shallow failures basically removed the weathered zone composed of tropical residual soils and rock fragments.

With regards to the angle of the natural scarp, no slope steeper than 40 degrees withstood the tremors and several flatter slopes failed as well.

6.4 Seismicity-related effects on slope failures

Relevant effects of the 1990 earthquake in Luzon were:

1) the lowering of cohesion and internal friction of soil materials. The reduction of intergranular bonds by the quake drastically lowered the amount of stability which is the excess of shear strength over shear stress. In rock slopes with a weathered zone the effect of ground shaking was selective. The process of regolith formation generates a weathering front which separates materials with different strength and hydrological properties, namely surficial soils and mother rock. Earth shaking weakened the intergranular cohesion-related bonds and grain-to-grain friction in the surface unconsolidated materials lowering their strength properties and consequently inducing sliding along the weathering front plane. Whenever the shaking intensity was too low to trigger slope failure, it loosened surface soils favoring water penetration. Hence, many dry slopes which withstood the 1990 quake failed during the seasonal rains.

2) the horizontal acceleration caused by the quake led to an increase in shear stress and the consequent failure of numerous slopes. Again, physical separation and different technical properties of the regolith and mother rock played an essential role with their different response potential.

During the quake slopes were mostly dry since the rainy season had not yet started, thus, failures induced by liquefaction did not occur. The distribution, orientation and pattern of faults also interacted with the geology and geometry of the landscape locally enhancing ground shaking intensities and, thus, the occurrence and dimensions of slides.

A remarkable effect of ground shaking in Luzon regarded slopes on bedded rock formations. Joints and fissures were considerably widened, new fractures were initiated and friction properties along major discontinuity planes decreased. Impressive structural failures occurred along Halsema and Kennon roads, north and south of Baguio respectively, due to sliding along the daylighting planes.

6.5 Shallow and deep-seated slides

The extensive failure of slopes in Luzon can be broadly divided into two groups: shallow-seated landslides involving the removal of the thin alteration cover and deep-seated landslides in which part of the bedrock with its weathered zone moves down the hillside.

The first group is typical of moderate to steep slopes with limited differences in elevation (Fig. 6.2). The height of the slope has no influence, failure being basically controlled by the slope angle. Sliding occurs along the contact plane between the 0.5 to 2 m thick weathered zone and the bedrock.

The joint pattern of the mother rock is marginally involved in slides of this type. Shallow-seated slides occurred extensively in the Caraballo Mountains (Dalton Pass Road) and in large areas of the Central Cordillera.

In the second group, deep-seated landslides, the failure involved the weathered zone and the bedrock (Fig. 6.3). Critical slope height, in general, is a major controlling factor together with the joint pattern, probably combined with the degree of weathering of the bedrock. Numerous slope failures of this type were observed in Cordillera Central and a few cases in the Caraballo Mountains.

Figure 6.4 shows in sequence the ridge opposite the slope along which the road leading from Agoo to Baguio (Marcos Highway) is located. The two pictures clearly illustrate that the steep slope has failed for kilometers. Debris slides, debris flows and avalanches, rock falls and rock slides in this area were associated with large-scale accelerated erosion.

Of the types of movement defined in the Varnes (1978) classification, slides were dominant, followed by flows and to a lesser extent by falls and complex types. Flows were abundant in the Cordillera Central but limited in the Caraballo Mountains. Slides were mostly translational, but cases of rotational types were also observed.

Fig. 6.2 – Hilly landforms of the Caraballo Mountains with the typical shallow-seated landslides (debris slides) along the Dalton Pass Road (top). Shallow debris slides near the road leading to Ambuklao Dam (Cordillera Central), NE of Baguio (bottom).

Fig. 6.3 – Huge deep-seated complex landslide (debris avalanche and rock slide) in the Caraballo Mountains. The failure involved the weathered zone and the bedrock.

Fig. 6.4 – Generalized failure of very steep slopes across the valley with debris flows, debris slides, debris avalanches, and many rock flows and some rock slides, seen from the Marcos Highway near Baguio. The two pictures are in sequence.

6

The slide materials included large quantities of rock fragments, which is an indication of the presence of mainly young weathered soils in the Luzon mountains. Coarse granular material was predominant in the deeply incised valleys in the Cordillera Central. In the high-topographic contrast area of the Caraballo Mountains, along Dalton Pass, some huge earth slides occurred (Fig. 6.11, top left). A few slides near Santa Fe' (Fig. 6.1) were characterized by a whitish color due to the failure of the coarse-graded quartz blanket covering granodiorites.

The abundant material which accumulated on the footslopes, was systematically eroded and transported by the heavy rains during the monsoon season.

6.6 Landscape evolution and environmental impact

In mountainous regions quakes can easily generate landslides, produce new cracks which can later develop into slope failures, and loosen surface materials, thus enhancing the erosive action of rain and flowing water. The July 1990 earthquake in this respect represents a major step in the evolution of the landscape in the Cordillera and Caraballo Mountains and an example of seismically induced enveronmental impact on a regional scale. In both ranges slides marked the landscape from the lowest elevations up to the farthest visible crests.

It is likely that a Magnitude comparable to that of the 1990 Luzon earthquake and intensities of VI to VIII on the Rossi-Forel scale were also reached in this area during previous strong tremors. The quakes of 1645, 1796 and 1892, with epicenters along the Philippine Fault segment between Dingalan Bay and the Gulf of Lingayen (Fig. 4.9), probably had the potential for producing an environmental impact similar to that induced by the July 1990 ground shaking.

According to Spanish chroniclers these events were highly destructive and caused considerable property damage. However, no systematic data were provided on what is now called the environmental impact. There are three reasons for this, a) Mountain Provinces were still widely covered by primeval forest, which attenuated slope instability and most certainly enabled disaggregated slope materials to progressively recover part of their original shear strength, b) there were very few access roads and the population was smaller and concentrated in coastal areas (thus, the information available mainly concerned inhabited flatlands), c) human action on vegetation and topography was minimal. The July 1990 quake, by contrast, and the following rainy season hit a quite, although not uniformly, populated region which had undergone deforestation and was widely farmed.

An active erosion cycle, in general, affects Luzon mountains due to the combination of heavy rains, tectonic uplift, river rejuvenation and deforestation. Fluvial downcutting is also very active and a marked slope retreat can be attributed to the combined effects of quakes and erosion. The erosive power of streams was evident for instance along Dalton Pass Road at the end of the 1990 rainy season. River sediments, transported and deposited between San Josè and Dalton Pass (Fig. 4.11), after the quake and seasonal rains, were deeply re-eroded down to a depth of 1 to 3 m.

The scars induced in Luzon mountains will probably remain visible for decades since deforestation and present erosion rate prevent rapid growth of a new vegetation cover. This will have an adverse effect on the fauna in general, increase runoff and reduce water seepage which is fundamental for the recharge of aquifers. The huge landslide-related impact already assessed in the latter part of 1990, was followed by additional devastation due to rain-induced slides and flow of sediments down rivers during the monsoon season of 1991.

6.7 Casualties, property and infrastructure damage

6.7.1 Death toll

Four hundred fifty lives were lost due to seismically-induced landslides and an additional forty casualties were reported due to later rainfall-triggered slope failures. In some of the most affected areas and remote zones landslides hampered search and rescue (SAR). The repeated traffic interruptions along

Dalton Pass Road due to rain-triggered slides not only damaged the economy but also delayed efforts by the SAR teams to provide medical supplies and help to remote villages.

6.7.2 Damage to property and roads

Property damage occurred in villages and on cultivated lands close to landslide prone areas. A number of small villages were reportedly abandoned or evacuated because of landslide hazard. People living along Dalton Pass Road, in particular, had to cope with repeated interruptions of traffic. There were over 100 small to medium sized slides and around 25 large slides, with an estimated overall volume of about 2 million cubic meters of debris along the 40 kilometers of the Dalton Pass Road between San Josè and Santa Fe' (Fig. 4.11).

Fig. 6.5 – *Debris slides along the Dalton Pass Road. Pavement, side slopes, drainage and often the entire road platform were badly damaged. Fissures induced by the quake affected the rigid pavement (top) and the shoulder (bottom), often with irreparable damage of the road surface. (Courtesy of DPWH and Katahira).*

A huge slide accompanied the earthquake in Bateria (Ligaya, Nueva Ecija Province), near the southern terminus of the major ground rupture (Wieczorek et al., 1990) close to Gabaldon (Fig 6.1). Along a very gentle slope angled at only a few degrees a huge quantity of sandy gravel and some boulders started to move slowly during the quake and traversed several hundred meters in the form of a semi-liquid mass. The translational movement was favored by the water-saturated condition of the mass.

Landslides contributed largely to structural damage in Baguio City. The airport runway and terminal, city buildings and houses suffered various types of damage due to soil movement and associated cracks. In a number of cases, however, rather than sliding, the soil underneath foundations was compacted by the tremors thus initiating foundation failure. According to Wieczorek (1990), landslides in Trinidad Valley (Baguio) swept some homes downslope.

Roads were affected by an enormous number of shallow debris slides (Fig. 6.5) and some rock falls, with consequent damage to cut slopes, retaining walls, side drainage, pipes and culverts, as well as to road surface.

Marcos Highway (Fig. 4.11) between Agoo and Baguio City was closed to traffic for some weeks after the quake. The most impressive landslide zone on this road is located a few kilometers before Baguio where very steep slopes (Fig. 6.4) along a deeply incised river valley dominate the landscape. In this case the entire slope was affected over several km.

In addition to shallow slides there were also deep joint-controlled plane failures on the Halsema and Kennon Roads (Fig. 4.11), respectively leading north and south of Baguio. The Naguillian Road, N-E of Baguio, also suffered damage from slope failures, while the road connecting Bagabag to Banawe (the village with the famous rice terraces) was closed for several weeks due to numerous landslides (Fig. 4.11 top).

The gravel road leading to Ambuklao Dam, northeast of Baguio, did not suffer major damage, but the surrounding landscape close to the damsite was badly affected by numerous debris slides. As in the case of Caraballo Mountains, contiguous surficial slope failures marked the landscape everywhere. A considerable volume of materials reworked by the monsoon rains filled up the reservoir of Ambuklao Dam in a few weeks, thus ending the electricity production.

Equally consistent was the downward movement of sediments along tributaries and major river beds during the August through October rainy season. The mass movements, which included rock fragments, plants, trees and debris, blocked or damaged bridges and drainage structures. After a month sediments transported along the natural drainage system reached the foothills and then started spreading out in the floodplains.

This natural movement of sediments endangered some bridge structures which had safely withstood the tremor. Countless uprooted trees blocked the concrete bridge in Puncan (along the Dalton Pass Road, Fig. 4.22), where the river bed had been raised by an unprecedented accumulation of sediments, which had also buried rice fields upstream. The Talavera and Digdig rivers along Dalton Pass Road (Fig. 4.11), were also choked by a major accumulation of debris. The sediments blocking the natural drainage in some cases caused rivers to overflow their banks.

6.8 Landslides in the Cordillera Central and Caraballo Mountains

6.8.1 General

The classification of landslides proposed by Varnes (1978) has been adopted (Fig. 6.6) top and terms are based primarily on the type of movement and secondarily on the type of material. The figure in brackets, which follows each landslide class in the next sections, represents a percentage of total landslides by type of material. The value is an indicative estimate derived from visual inspection of a number of sites covering about 280 sq. km in Southern Cordillera and 150 sq. km in Caraballo Range. Considering that the latter range is smaller and morphologically more uniform, the percentages of slope movements are comparatively more representative for the Caraballo Mountains. The most common types of landslides are illustrated in the lower part of Figure 6.6.

6.8.2 Types of slope failure in the Southern Cordillera

The distribution of landslides in shown in Fig. 6.1 which also includes isoseismal lines. Slide types in the Southern Cordillera mountains were classified as follows:

a) debris slides (33%) and debris flows (14%). Debris slides in hilly landforms are shown in Figures 6.2 (bottom) and 6.7, while Figures 6.8 and 6.9 illustrate debris flows and some debris slides in shattered rocks forming steep-sided slopes near Baguio.

b) rock flows (6%) and debris avalanches (7%). A number of rock flows and debris avalanches associated with other types and accelerated erosion occurred on steep slopes in the Central Cordillera (Fig. 6.4).

TYPE OF MOVEMENT			TYPE OF MATERIAL		
			BEDROCK	ENGINEERING SOILS	
				Predominantly coarse	Predominantly fine
FALLS			Rock fall	Debris fall	Earth fall
TOPPLES			Rock topple	Debris topple	Earth topple
SLIDES	ROTATIONAL	FEW UNITS	Rock slump	Debris slump	Earth slump
	TRANSLA-TIONAL	MANY UNITS	Rock block slide	Debris block slide	Earth block slide
			Rock slide	Debris slide	
LATERAL SPREADS			Rock spread	Debris spread	Earth spread
FLOWS			Rock flow (Deep creep)	Debris flow / Debris avalanche (soil creep)	Earth flow
COMPLEX			Combination of two or more principal types of movement		

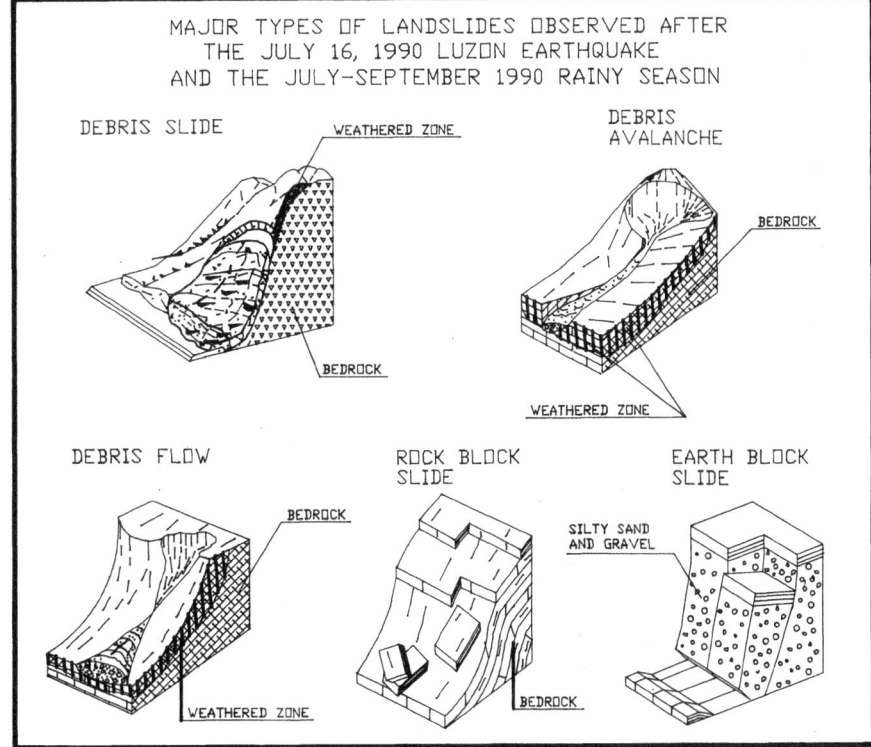

MAJOR TYPES OF LANDSLIDES OBSERVED AFTER THE JULY 16, 1990 LUZON EARTHQUAKE AND THE JULY–SEPTEMBER 1990 RAINY SEASON

Fig. 6.6 – *Varnes classification (1978) of slope movements (top) and most common types of slope instability phenomena in Luzon (bottom) triggered by the earthquake and the rainy season which followed soon after.*

85

Fig. 6.7 – *Shallow-seated landslides (mainly debris slides) along the road leading to Ambuklao damsite.*

Fig. 6.8 – *Debris flows, debris slides and accelerated erosion along the steep slopes of the Marcos Highway near Baguio.*

Fig. 6.9 – *Debris flows, some debris slides and accelerated erosion along the Marcos Highway, near Baguio. The deep-seated slide on the far right is interpreted as a complex type of movement combining falling, sliding and flowing.*

c) rock slides (10%). Slope failure controlled by joints and other discontinuities of the rock mass occurred along Halsema and Kennon Roads.

d) rock falls (7%). This type of instability was common on steep slopes but the quantity of material mobilized was usually minimal.

e) rock block slides (5%). Typical cases of plane failure along bedding planes occurred along the Halsema and Kennon Roads. This joint-controlled type of slide involving the bedrock is quite common in folded sedimentary rocks. Figure 6.10 shows an example of sliding along inclined sandstone layers near Baguio (Halsema Road). The angle of the natural slope coincided in this case with the dip of bedding planes. The construction of the road, however, contributed to the sliding by bringing the bedding planes to daylight through the slope face.

f) complex (9%). Figure 6.9 (Marcos Highway near Baguio) shows a complex slide on the right due to the combination of rock fall, rock flow and some cutting through the intact rock material.

g) others (9%).

6.8.3 Types of slope failure in the Caraballo Mountains

Over 400 slides of different types were counted along the Dalton Pass Road (Maharlika Highway); construction, repair and drainage works in this case heavily influenced the behavior of slopes during the quake. More uniformity, instead, was observed in the slope failures at a distance from the road and generally in the southern part of Caraballo Mountains.

Fig. 6.10 – *Plane failure along the Halsema Road north of Baguio in sandstone rocks. Beyond the ground shaking, the two factors that affected the pulling force were the angle of slope and the daylighting of the bedding planes.*

Fig. 6.11 – *Typical area of debris and earth slides along very steep slopes near Dalton Pass (top left). Earth block slides induced by undercutting along Digdig river (top right). Debris slides along the Digdig River (near Capintalan) and the huge fan created during the 1990 rainy season (bottom).*

Slope failures were classified as follows:

a) debris slides (55%). Shallow debris slides are dominant and their main characteristics are (Figs. 6.2 top and 6.11 bottom):

  - moderate to steep slopes (35 - 55 degrees). The critical slope angle above which failure occurred is in the 35 to 38 degree range, depending on the type of regolith, bedrock, and the quantity and type of fines; slope height is usually moderate and the topography is predominantly hilly. The thickness of the slides is between 0.5 and 2.0 m.

  - the most common type of slide material is a light-brown mixture of partly weathered angular rock fragments, ranging from 2 to 15 cm in size, sand and silt; some large debris slides with a predominant light-gray fine matrix occurred south of the Dalton Pass (Fig. 6.11 top left);

  - the detachment area is variously shaped depending on local geological and topographic conditions, but generally marked by a sharp line.

b) debris flows (15%) and debris avalanches (3%). Some occurrences were observed along the Dalton Pass Road. The huge debris avalanche (and rock slide) in Figure 6.3 is thought to have been reactivated by the quake.

c) rock slides (4%). The slides of instability was not very common. A number of cases vere observed along the road leading from Sta. Fe to Imugan (Fig. 4.11).

d) earth block slides (4%). The slides of this type along undercut banks of streams were mainly activated by the rains which followed soon after the quake. Examples of downcutting along the Digdig river are shown in Figure 6.11 (top right). The material is composed mainly of sandy gravel.

e) rock fall (3%). Very few rock falls were observed along the Dalton Pass Road.

f) complex (7%). A number of slides combining two or more types of movement were observed along the Dalton Pass Road (Fig. 6.3).

g) others (9%).

The slide material reworked and transported during the July-November 1990 rainy season created spectacular alluvial fans (Fig. 6.11 top right and bottom) near the Dalton Pass road.

CHAPTER 7

PRE-EARTHQUAKE SEISMICITY
AND THE 1990-91 AFTERSHOCK SWARM

7.1 The July 1990 earthquake and aftershocks

During the quake the region including the Sierra Madre Range and Cagayan Valley Basin was displaced northwestward, the part of Luzon south and west of the ruptured segment being subject to compression.

The accumulation of stresses mainly regarded the area southwest and west of the ground rupture, and north along its probable underground extension. As a result, a number of bridges, barriers and undulations along faults in the Central Valley basement and the Cordillera were progressively sheared off during the six months after the main quake. This caused numerous small and a few medium-intensity tremors throughout the central and north-western parts of the island, but with little damage.

The aftershock swarm was interpreted as a regional subsurface block rearrangement within the area of the July 1990 quake. The reorganization of the basement in Central Luzon, progressively affected the shallower crustal zone where confining pressure was lower, squeezing fractures and causing the intrusion of molten rock into chimney channels of volcanoes in the area.

The formation of foci, which mainly clustered southwest and west of the ground rupture zone, gradually shifted in a northerly direction, more or less following the same trend as the progressive ground rupture on July 1990. The depth of aftershock hypocenters progressively decreased a few days after the main quake.

On July 16, 1990, 32 shocks, most between 4.7 and 5.7 Magnitude, occurred worldwide within a period from 2 hours before to 16 hours after the Luzon earthquake (Table 7.1). Four were in Japan, two in Alaska, two in Taiwan, and one each in Iran, Guatemala and Chile, while the remaining 21, located in the Philippines, concentrated in the 14 hours after the Luzon quake. Twenty of these shocks affected Luzon and one Mindanao. This clearly shows the persistent high seismicity of Luzon associated with the strike-slip motion along the Philippine Fault.

7.2 Seismicity trends between 1985 and 1991 in the Philippines

The distribution of seismicity before 1990 was relevant to the July 16 earthquake and the associated aftershock swarm. Figure 7.1 shows the location of epicenters within the Archipelago during the period 1985-1991.

Seismic activity occurred during 1985 with clusters of epicenters in Baguio and Mindanao. The northwestern coast of Luzon was marked by scattered epicenters. The Archipelago was very quiet during 1986, with comparatively few epicenters and limited clustering. During 1987 seismic activity increased again with the formation of a cluster between Leyte and the northeastern part of Mindanao, along the

TABLE 7.1 - Location and Magnitude of epicenters worldwide from 2 hours before to 16 hours after the July 16, 1990 earthquake in Luzon

US. DEPARTMENT OF THE INTERIOR - GEOLOGICAL SURVEY - NEIC QUICK EPICENTER DETERMINATIONS

NO. 0-204
JUL 23, 1990

UTC TIME HRMNSEC	LAT	LONG	DEP	GS MB	MAGS Msz	SD	STA USED	REGION-COMMENTS
JUL 16								
051110.4p	32.36 N	142.40 E	33N	5.0		0.8	9	S OF HOHSHU, JAPAN
055628.3s	13.374N	91.162W	33N	4.7		1.4	20	NEAR COAST OF GUATEMALA
071225.4	24.276N	121.816E	33N	5.1	4.0	1.0	21	TAIWAN
072635.9	15.675N	121.257E	36	6.7	7.7	0.9	151	LUZON, FHILIPPINE

ISL. Mo=8.0*10**20 Nm (PPT). At least 862 people killed, more than 3,000 people injured and severe damage and landslides in the Baguio-Cabanatuan-Daguapan area. Large fissures were observed in the epicentral area. Damage also occurred in Bataan Province and at Manila. Felt (VII RF) in the Manila area, (VI RF) at Santa and (IV RF) at Callao Caves.

083533.9s	16.299N	120.871E	33N	5.4		1.4	28	LUZON, FHILIPPINE ISL
084358.7p	15.62 N	121.29 E	33N	4.9		1.6	7	LUZON, FHILIPPINE ISL
085023.4	16.479N	121.041E	33N	5.1		1.0	28	LUZON, FHILIPPINE ISL
091608.5s	16.407N	120.976E	33N	5.2		1.4	21	LUZON, FHILIPPINE ISL
092909.2	16.455N	120.400E	33N	5.7		1.0	47	LUZON, FHILIPPINE ISL
093516.2p	16.96 N	120.27 E	33N	4.9		0.7	8	LUZON, FHILIPPINE ISL
093923.7	16.498N	120.935E	33N	5.7		1.3	32	LUZON, FHILIPPINE ISL. Felt (II RF) at Callao Caves.
100814.4	36.266N	141.328E	33N	5.1		1.0	44	NEAR E COAST OF

HONSHU, JAPAN. Felt (III JMA) at Mito and (II JMA) at Choshi.

101257.4	9.215N	125.477E	33N	5.2		1.2	38	MINDANAO FHILIPPINE ISL
102424.7	16.467N	120.962E	33N	5.1		1.4	28	LUZON, FHILIPPINE ISL
121836.9	16.067N	121.01E	33N	4.8		0.7	11	LUZON, FHILIPPINE ISL
123510.7	39.344N	141.965E	61	4.9		0.8	22	HONSHU, JAPAN
130242.6s	15.883N	121.125E	33N	5.5		1.3	31	LUZON, FHILIPPINE ISL
130716.8s	59.152N	151.897W	118p			0.8	11	KENAI PEN, ALASKA. Felt (III) at Homer
133117.2	16.434N	120.320E	33N	5.6	5.4	1.2	37	LUZON, FHILIPPINE ISL
145136.2	32.520S	70.096W	106D	5.5		1.0	62	CHILE-ARG BDR REG.

Felt (VI) at Calera., Illapel, Los Molles and Salamanca; (V) at Cabildo, Combarbala, La Ligua, Los Andes, Valle Nevado and Vicuna; (IV) at Santiago and Valparaiso; (III) at La Serena and (II) at Talca, Chile. Also felt (IV) at Mensoza, Argentina.

150428.2	16.525N	120.248E	33N	5.5	4.5	1.2	32	LUZON, FHILIPPINE ISL
161934.3s	16.408N	120.366E	33N	5.5		1.2	33	LUZON, FHILIPPINE ISL
165338.5s	15.804N	121.305E	33N	4.7		1.2	7	LUZON, FHILIPPINE ISL
18037.3s	28.511N	56.979E	33N			0.2	7	SOUTHERN IRAN
191458.7	24.329N	121.813E	75a	5.5		1.0	50	TAIWAN. Felt at Taipei
194525.3s	16.437N	120.526E	33N	5.1		0.7	12	LUZON, FHILIPPINE ISL
201606.7s	16.650N	120.428E	33N	5.2		1.2	27	LUZON, FHILIPPINE ISL. Felt (III RF) at Santa
203124.7	17.518N	121.045E	33N	5.3		1.0	41	LUZON, FHILIPPINE ISL
204418.9	15.479N	121.302E	33N	5.2		1.1	20	LUZON, FHILIPPINE ISL
212621.4	16.322N	120.469E	33N	5.1		0.9	18	LUZON, FHILIPPINE ISL
224428.1n	60.012N	153.285W	145p			0.5	7	SOUTHERN ALASKA
233016.5s	36.751N	141.414E	33N			0.8	10	NEAR E COAST OF HONSHU, JAPAN

golden co usa 1990 JUL 23 12:12
From: 6780::NEIS::ALERT   23-JUL-1990 19:27:28.25
To: ISDRES::40302::BOLLET

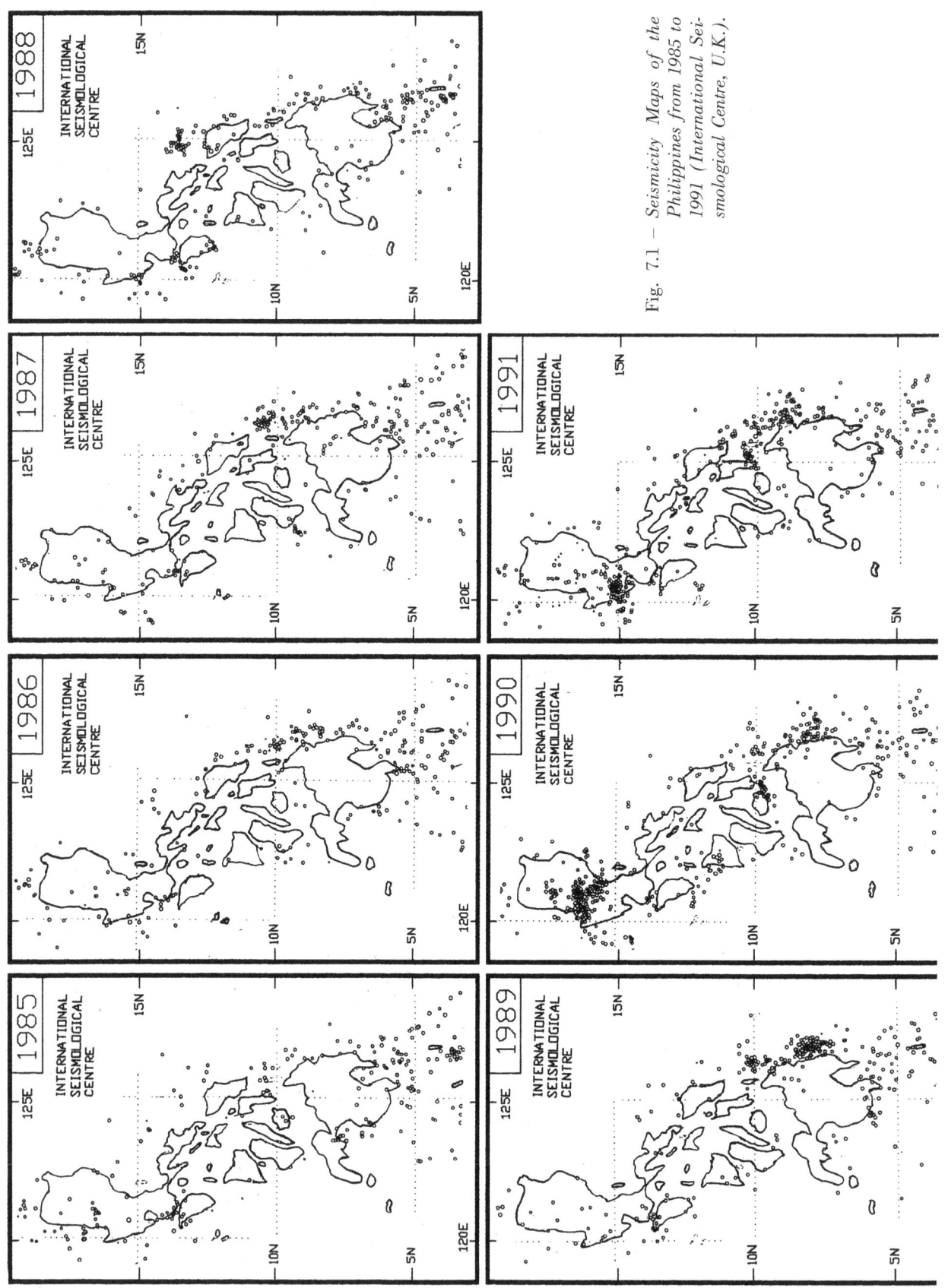

Fig. 7.1 – *Seismicity Maps of the Philippines from 1985 to 1991 (International Seismological Centre, U.K.).*

Philippine Trench. Activity continued during 1988 with a cluster east of Mayon Volcano (Legaspi). Almost no seismicity was recorded in Luzon during 1989, but two clusters formed east of Mindanao within the Philippine Trench zone. This year of relative calm in Luzon gave way to very high seismicity in mid-1990, with the formation of some major clusters in central and northern Luzon and some minor ones west and north of Mindanao Island. The abrupt change in the seismicity level started with the July 16, 1990 earthquake in Luzon.

The impressive difference between 1989 and 1990 shows that during the period of relative calm stress release must have been prevented by the presence of insuperable obstacles to movement along faults. The shearing off of these barriers required a long accumulation of stress which lasted during the whole of 1989 and half of 1990, ending with the July 16, 1990 events and numerous aftershocks during the same day and the next four days. Numerous epicenters clustered around Mt. Pinatubo in 1991.

7.3 Aftershock swarm

Figures 7.2 and 7.3 show the major features of the aftershock swarm, namely, a) the tendency of shocks to form clusters in specific areas, b) the northward trend in the formation of clusters, c) the importance of the Baguio cluster covering about 200 sq. km.

The area affected by the aftershock swarm extends for nearly 400 km, from Dingalan Bay to the northwestern tip of Luzon (Fig. 7.3). Cluster development from July 1990 through February 1991 indicates that a complex interaction occurred during this period, with contiguous blocks influencing one another and contributing to the formation of new temporary equilibrium conditions. The rate of subsurface block readjustment in Central Cordillera, based on aftershock occurrence, was critical from July 16 to July 20, 1990, both in terms of the number of tremors and the presence of some Magnitude 6 events; it then started to decrease. Figure 7.2 shows the progressive formation of clusters of epicenters between July 1990 and April 1991 while Figure 7.3 illustrates the position of clusters relative to the fault system in Luzon.

The formation of clusters (Fig. 7.3) can be summarized as follows:

a) the first, located in the southernmost part of the rupture, has few epicenters, which are broadly scattered between Gabaldon and Bongabon;

b) the second, is a slightly larger concentration of epicenters extending along and to the west of the ground rupture, between Digdig and Rizal;

c) the third, located in the Kayapa-Baguio-Lingayen Gulf area represents the largest concentration of epicenters. It broadly extends west of the ground rupture, with a decrease in the density of epicenters towards the Gulf of Lingayen. The shape and the area involved indicate a considerable activity associated with motion along major faults and sub-faults near and south of Baguio.

d) the fourth, located near Bengued City District (Abra Province), suggests that the subsurface slip induced by the July 1990 quake probably continued underground beyond Kayapa along the Abra River Fault. The progressively northern trend in seismic activity along the Digdig and Abra Faults suggests that the two faults actually form a single tectonic lineament.

e) the fifth is a minor concentration of epicenters in the area south of Pasaleng Gulf. It appeared only in 1991 (January through April), thus it is the last in the time sequence and the most northerly in terms of location. Its position suggests that the postulated underground slip motion (beyond the northern terminus of the July 1990 ground rupture) might have influenced block motion as far north as the area between Luzon and Taiwan.

The upward migration of hypocenters four days after the July 16, 1990 events is shown in Table 7.2, which covers the aftershock period July 16 - October 19, 1990 (Besana et al., 1990). Table 7.2 provides a clear indication of the depth and trend of subcrustal readjustment, based on the upward migration of hypocenters.

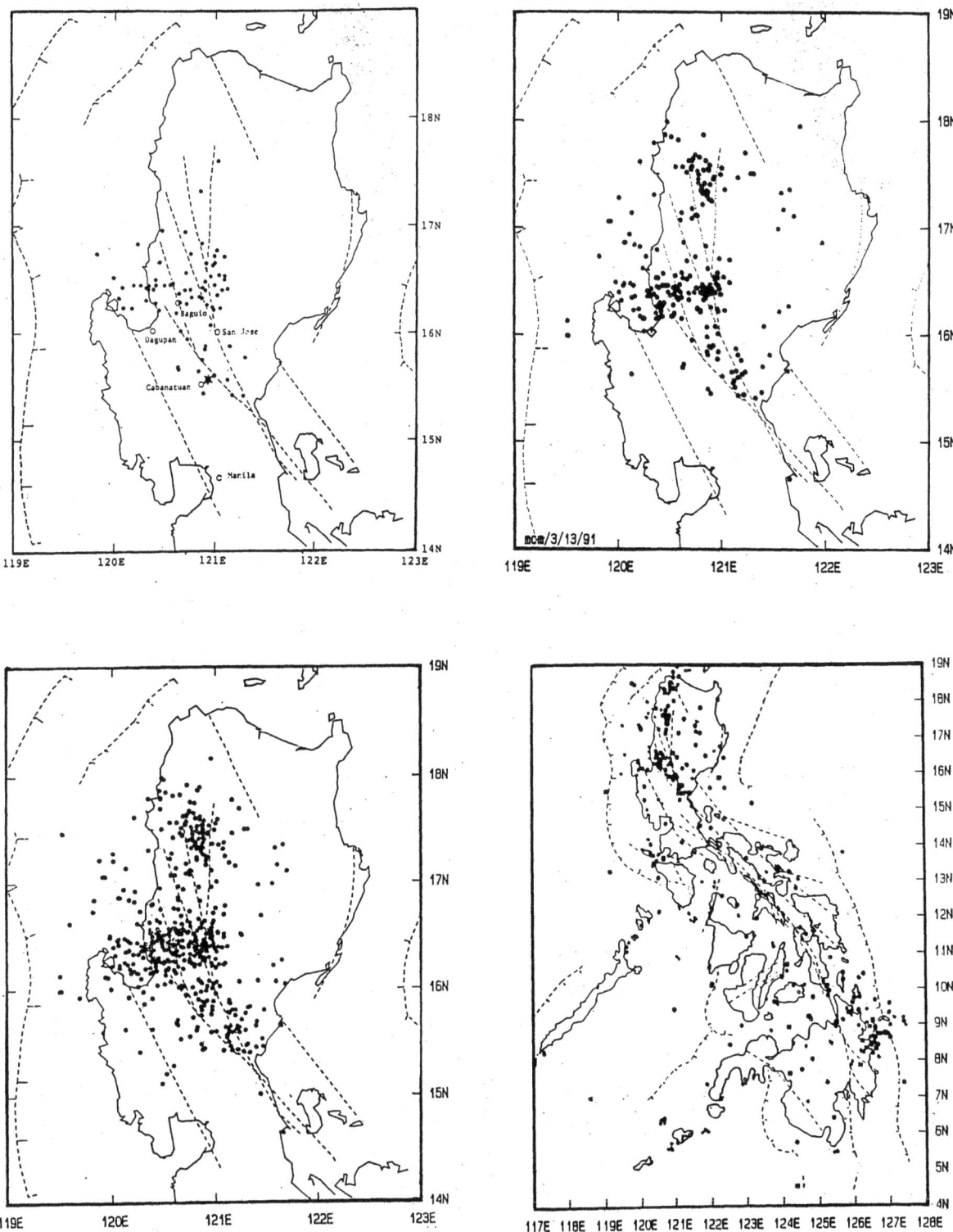

Fig. 7.2 – *Distribution of aftershocks within two weeks after the main July 16, 1990 shocks (top left), between July 1990 and January 1991 (top right), between July 1990 and February 1991 (bottom left), by Besana et al. 1990. Seismicity of the Philippines between January and April 1991 (bottom right), by PHIVOLCS.*

The five periods into which the aftershock sequence has been divided are based empirically on various factors such as the number of tremors per day, pauses of a few days between homogeneous after-shock episodes and magnitudes.

The first group (July 16-20), for instance, includes numerous shocks associated with the major events of July 16. Fifty-seven events in five days, with an average of nearly 12 shocks per day, represent the most critical episode with the majority of quakes at a depth of 32 to 33 km.

The second period, ending August 7, is characterized by a marked drop in frequency to an average of four shocks per day with the majority at depths between 1 and 10 km. Third, fourth and fifth episodes, separated by pauses of a few days almost entirely free from seismicity, are characterized by an average of 2 to 3 events per day, most of them at depths of 1 to 10 km.

A clear trend of depth of foci with time emerges: during the first five days 56% of hypocenters are located at 33 km depth and 37% between 1 and 23 km, while during the July 21 through October 19 period an average of 59% of hypocenters has already migrated to the 1 to 10 km depth range.

The definite upward migration of foci in the top 10-km crustal zone, suggests that barriers and jogs getting into a critical condition and then sheared off, were located at progressively shallower depths where confining pressure is lower.

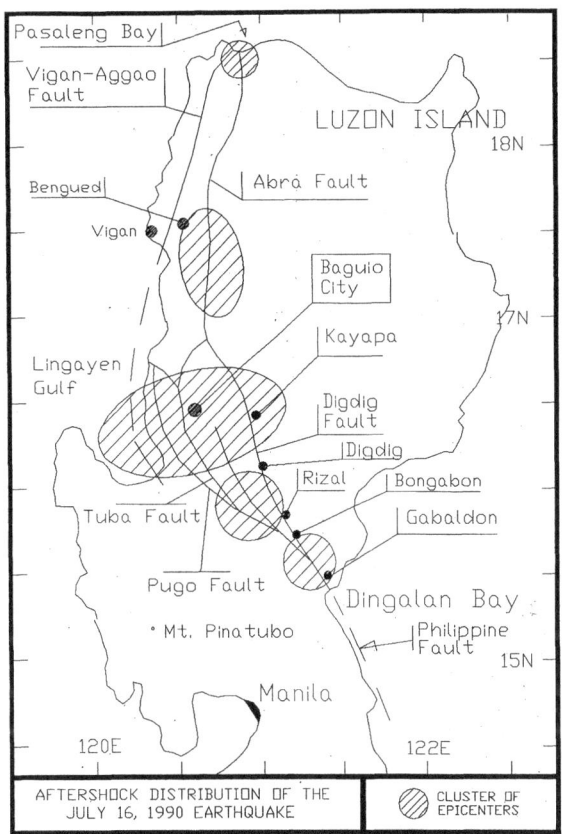

Fig. 7.3 – *Map of the clustering of aftershocks between July 1990 and May 1991.*

TABLR 7.2 - Depth distribution of hypocenters during the period July 16 - October 19, 1990							
Period	N. of Events	1-10 km	11-23 km	24-31 km	32-33 km	34-59 km	67-161 km
16-20/7	57	6 11%	15 26%		32 56%	1 2%	3 5%
21/7-7/8	71	38 53.5%	18 25%	5 7%	9 13%		1 1.5%
11/8-2/9	44	29 66%	7 16%	2 4.5%	1 2%	2 4.5%	3 7%
6-28/9	50	31 62%	16 32%	1 2%	1 2%		1 2%
3-19/10	45	25 56%	14 31%	3 7%		2 4%	1 2%

7.4 Concluding remarks

Some observations can be made regarding the July 16, 1990 events and the aftershock swarm:

- the compressional regime in western Luzon persisted long after the major and minor July 16, 1990 tremors. In addition to barriers sheared off on July 16, other obstacles and undulations were progressively eliminated as soon as their stress condition became critical.

- the upward migration of hypocenters from 33 to between 1 and 10 km depth is indicative of block readjustment having been virtually completed in a few days at the higher stress levels at 32-33 km depth.

- the location and timing of the Bengued and Pasaleng Gulf aftershock clusters suggest that the ground rupture proceeded underground northward along the Abra Fault;

- the seismic quiescence in 1989 (and probably part of 1990) in contrast to the high level seismicity of July-December 1990 suggests that seismicity level variations may be significant. An investigation on «seismic gaps» and their implications could provide clues to future strong earthquakes.

- the trend and timing of the crustal readjustment in Luzon (induced by the July 1990 earthquake) is thought to have caused the intrusion of molten rock into the fractures below Mount Pinatubo and the warming up of Mount Taal (south of Manila) during 1991.

CHAPTER 8

THE ERUPTION OF MOUNT PINATUBO IN 1991

8.1 Introduction

Mount Pinatubo (1,745 m altitude before the collapse of the crater in 1991) is located in the southern part of the Zambales Range, west of the Central Plain. As the highest Quaternary cone of Central Luzon, Pinatubo belongs to the Bataan Lineament, a sequence of calc-alkaline edifices, forming part of a volcanic arc several hundred km long. Figure 8.1 provides an overview of the physiography of the region with four volcanoes, namely the active Mt. Pinatubo and three dead cones nearby. Two, Mt. Natib and Mt. Mariveles, are massive and composite edifices forming the Bataan peninsula, west of Metro Manila; the third, Mt. Arayat, is an isolated cone in the Central Valley some 35 km east of Pinatubo. Most of the huge alluvial plain extending N and S of Mt. Arayat is drained by the Pampanga River, its affluents and an intricate network of minor streams and channels, all flowing into Manila Bay.

During the eruption of Mt. Pinatubo, between June 12 and 15, 1991, the volcano spewed out several cubic kilometers of ejecta, most in the form of ash. Heavy pyroclastics were deposited in abundance near the crater, while steam clouds carrying huge quantities of ash rose to a height of thousands of meters.

Part of the dust forming a mushroom cloud was directed westward by typhoon winds, and part of it was deposited over the volcanic landscape, the thickness of the cover decreasing with distance from the crater. As a result of the eruption a 10,000-square kilometer ashfall blanketed Central Luzon provinces bringing death and devastation. The part of the ash cloud driven westward soon approached the coast of Vietnam and within a month it had circled the globe, reaching the Hawaiian archipelago by July (National Geographic, May 1992).

Further havoc was wrought by the destructive power of hot and cold mudflows, or primary and secondary lahars (Indonesian), the slurry-like streams of water-saturated volcanic ejecta originating from the slopes of Pinatubo during and after the eruption. During the major explosive phase of June 1991 the monsoon downpours saturated and then mobilized the ash cover thus starting cold lahars; the rains also enhanced the devastating potential of primary lahars which had originated directly from the high temperature mixture of ejected ash and steam. Through the numerous streams and tributaries dissecting the volcanic landscape, the gray mixture flowed down the natural drainage system, finally reaching the flatlands, engulfing villages, filling depressions and causing huge devastation.

Figure 8.2 illustrates the hazard zones with the circles indicating the radial distance from the crater, and the municipalities affected by the ashfall or/and by mudflows.

8.2 Geo-tectonic setting

The geology and tectonics of Central Luzon were described in Chapter 3. More detailed information on the volcanism of the region is provided here. The Philippine Archipelago is characterized by the presence of volcanic arcs (linear sequences of volcanoes) essentially parallel to subduction zones. The descent of the plates to a depth of several dozen kilometers under Luzon causes them to melt, while their collision produces block movements along faults and thus, earthquakes, as well as the mobilization

Fig. 8.1 – *Physiographic features of Mount Pinatubo and nearby volcanoes.*

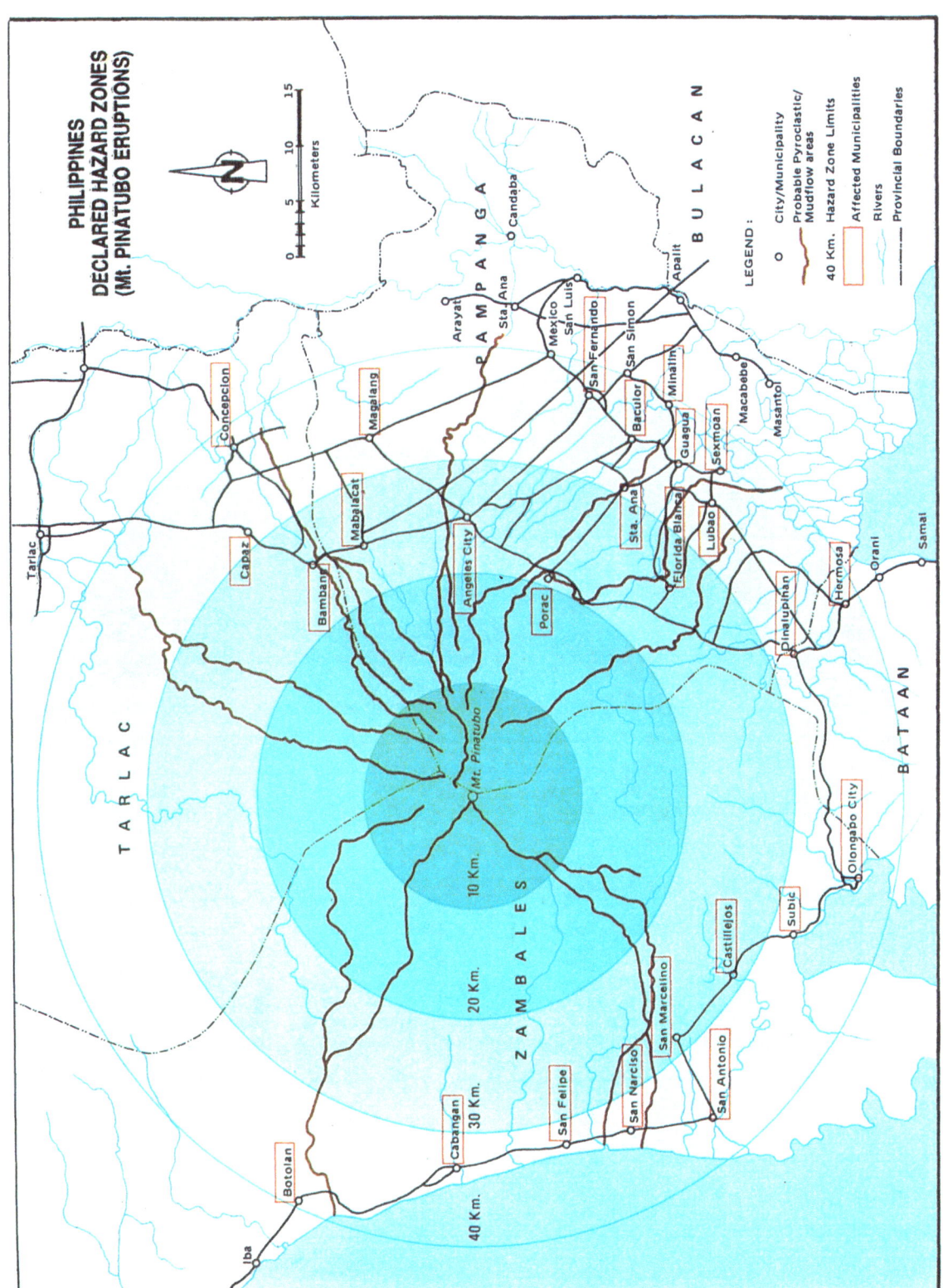

Fig. 8.2 – *Map of declared hazard zones due to lahars activated by monsoon rains (ADB, 1991b).*

99

of the molten rock formed near Benioff Zones. Under these conditions the molten material can find its way upward through fractures to the ground surface often producing powerful eruptions.

Of the four arcs that occur in the Philippines along mobile subduction zones, the West Luzon Volcanic Belt is of particular importance; Mount Pinatubo is situated in the northern part of this arc, known as the Bataan Lineament.

The mountain chain west of the Central Plain, commonly known as the Zambales Range, is actually composed of two different and partly overlapping sub-units. The Zambales Mountains in the west consist of an ophiolitic N-S oriented rock complex about 130 kilometers in length.

The Bataan orogen is a line of recent volcanoes bordering the ophiolitic sequence on the inner side. From Mount Pinatubo, which is the major, highest and northernmost composite structure, the line of volcanoes strikes southward for about 320 kilometers, at an average distance of 100 km from the Manila Trench. The Bataan Lineament also includes the extinct volcanoes Natib and Mariveles on Bataan Peninsula. The bold dotted line in Figure 8.3 shows the location of the Lineament, which has 27 vents and includes huge calderas, strato-volcanoes, isolated and composite cones, and minor vents (Wolfe et al., 1983). Figure 8.4, an enlargement of the box in Figure 8.3, illustrates the tectonic features of this area.

The Bataan Lineament extends southward beyond the Verde Island Strait. Its products are calc-alkaline and its activity was mainly explosive (see also Chapter 3), the arc being associated with subduction along the Manila Trench. The lineament is still partly active, as shown by recent events of Pinatubo and by the more than ten eruptions of Mt. Taal (Fig. 8.4) since 1965. The 1754 and 1911 phreatomagmatic eruptions of Taal are among the most catastrophic in the history of the Philippines. The former event significantly altered the morphology of Taal's caldera (Hargrove, 1991) by the closure of its wide channel to the sea with a barrier of pyroclastics and the consequent gradual change from a saltwater bay in the caldera into the present fresh-water lake.

Mount Pinatubo original andesitic products later became dacitic, while the most recent eruption, before the 1991 events, consisted mainly of plagioclase-rich tuffs (ignimbrite). The major activity of the lineament is believed to have taken place between the lower Miocene and the Pleistocene (Appendix A). Figure 8.5 shows some recent volcanic eruptions along the margins of the Pacific Plate region. Five of the events between 1991 and 1993, are located along the volcanic arc stretching from the Philippines to Kamchatka peninsula.

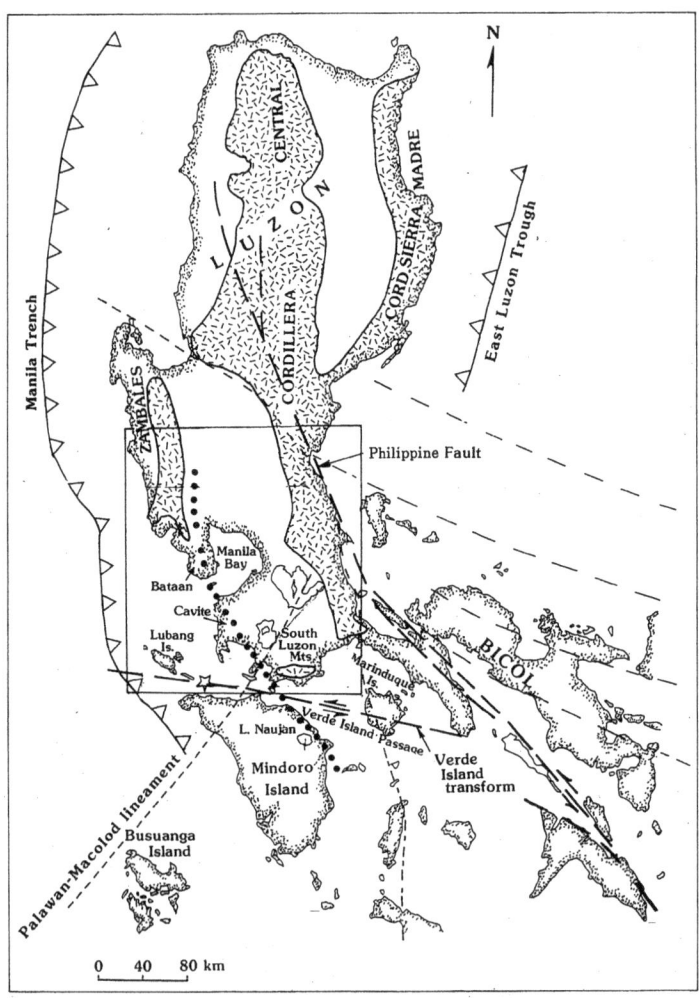

Fig. 8.3 — *Generalized Map of Luzon with the mountain ranges in dash pattern (Wolfe and Self, 1983). The dotted line indicates the location of the Bataan lineament hosting Mt. Pinatubo. Reprinted by permission of the American Geophysical Union.*

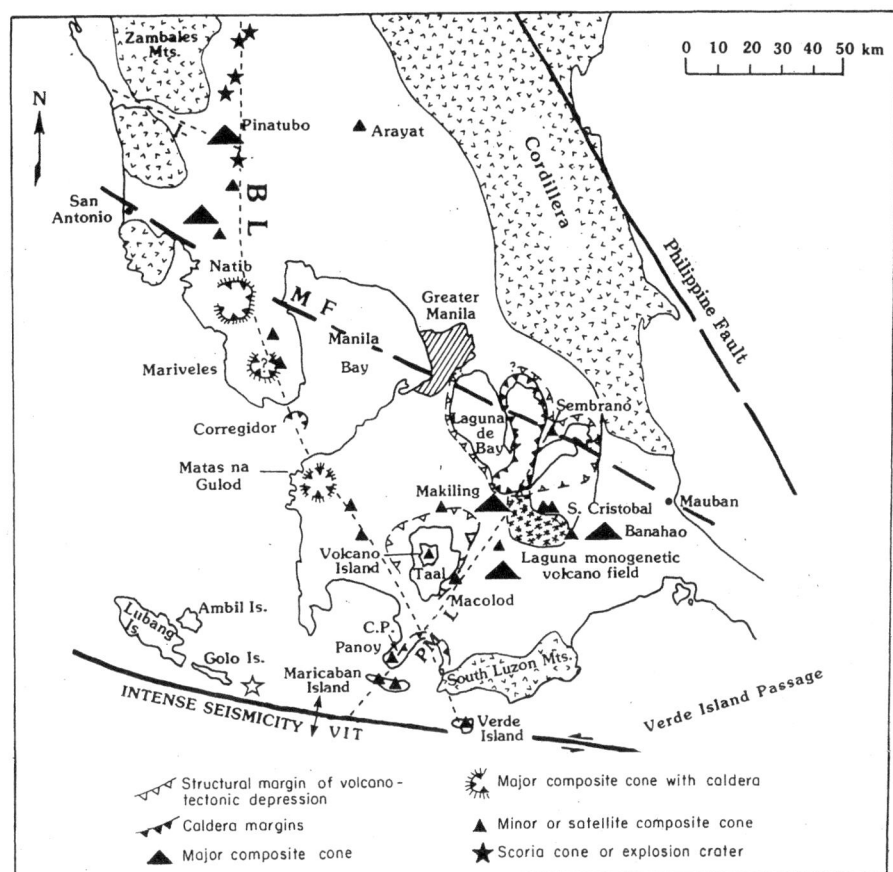

Fig. 8.4 – *Detail of box in Fig. 8.3. Location of the Neogene volcanoes and the major tectonic lineaments of Central Luzon (VIT, Verde Island Transform Fault; MF, Manila Fault; PML, Palawan-Macolod Lineament, BL, Bataan lineament), by Wolfe and Self, 1983. Reprinted by permission of the American Geophysical Union.*

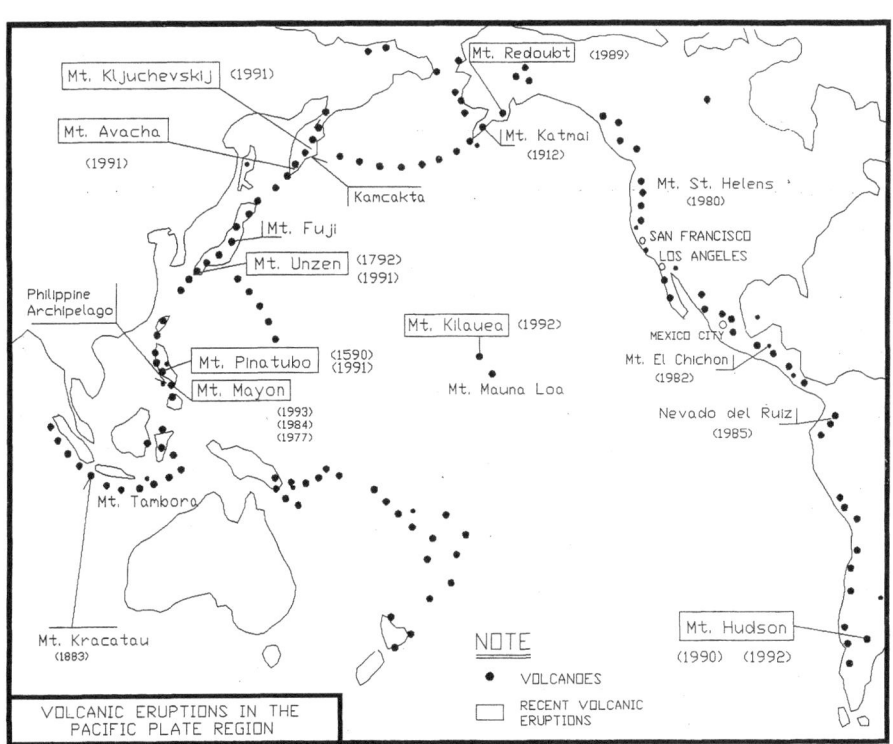

Fig. 8.5 – *Overview of eruptions by the major volcanoes in the Pacific Region during the period 1991-93.*

The June 1991 eruption of Pinatubo is attributed to the rearrangement of subsurface blocks which followed the July 16, 1990 Luzon quake. The block readjustment caused intrusion of molten rock along fractures and conduits upward into the cone.

8.3 Mount Pinatubo eruption

8.3.1 Pre-eruption signs and major explosive episodes

According to Punongbayan, in August 1990, a month after the July 16 earthquake, roars, ground cracking and a higher steaming activity of the thermal area of Pinatubo were reported to PHIVOLCS by the indigenous Aetas living around the volcano. The response team attributed the aforementioned signs to landslides initiated by the numerous aftershocks which followed the quake and by the monsoon rains, thus ignoring the possibility of an approaching eruption.

More consistent signs of Mount Pinatubo awakening after about 400 years of quiescence (according to a recent Japanese dating) arrived at the beginning of April 1991, in the form of steam clouds and ash emissions. Accompanying seismicity soon indicated movements underneath the crater and the possibility that they could turn into an eruption. Shallow-seated seismic activity, smoke mixed with ash and the emergence of a dome on June 8 persuaded government authorities to start the evacuation of the numerous members of the Aetas tribe (Para. 8.10) living on and around the volcano. In the meantime sizable carbon dioxide and sulphur dioxide emissions indicated that magma was rising along the volcanic conduit and nearby fractures (Mount Pinatubo, 1991). On June 9, moderate explosions resulted in ash and smoke reaching Clark Air Base, nearly 20 km southeast of the crater. Volcanic ejecta soon began falling at ever greater distances, gradually approaching the more densely populated areas.

A powerful explosion from Mount Pinatubo, 8.51 in the morning on June 12, 1991 created a 20 km high ash cloud which continued to ascend steadily. According to PHIVOLCS reports, the cloud moved in a northwesterly direction and obscured the sky; high frequency tremors continuously rocked the land with ever increasing intensity, while the magma dome started growing. During the day numerous explosions were reported and abundant ashfalls and ejecta reached the cities of Olongapo and Angeles, as well as villages within 20 km from the crater. As the dome became more and more dangerous, thousands living around the volcano started moving out of the hazardous areas, while ashfalls had already blanketed hundreds of square kilometers.

Another strong explosion took place on June 13 and ash fell as far away as the Zambales, Tarlac and Pampanga provinces (Fig. 8.2). Pyroclastic flows also started moving along incised valleys around the volcano reaching considerable distances, while the ash cloud had already spread over the South China Sea. On June 14 five eruptions rocked the volcano spewing gases and ash, hot lahars began cascading from near the crater, and were made more mobile by heavy rain. On June 15th, 16 explosions occurred; ejecta were spewed everywhere and numerous explosions took place in a crescendo of seismic activity until the upper edifice of the volcano collapsed forming a new crater 2 by 3 km wide (Mount Pinatubo, 1991).

Figure 8.6 shows the volcano seen from Abacan River near Angeles (top left), a few weeks after the paroxysm, and the ash cloud as of June 1991 (top right). The bottom figure shows the northwestern border of the crater after the collapse (picture taken from PHIVOLCS Observatory, upstream of O'Donnel River). Numerous hot lahars, depending on the volume of steam-ash emissions, were generated during the critical phase (isolated smoke emissions and small hot lahars were still active in September 1991).

By June 18 Manila, 110 km SE of the crater, was darkened in daytime and blanketed by half a centimeter of ash. The deposition extended far to the west and south of the volcano whitening the South China Sea waters for days. Clark Air Base, Angeles City and villages east, south and west of the crater were heavily covered by ash. There was no panic but several thousands of refugees had already abandoned the slopes around the volcano to be relocated in safer areas. During the second half of June ash clouds passed over the South China Sea and reached Vietnam and Cambodia.

By the end of the month over 500 victims were feared, numerous injured and a number of missing, and more than a million people had been affected by the devastation. Despite the risk, however, some Aetas refused to abandon their land around the volcano.

Fig. 8.6 – Mount Pinatubo seen from the SE with the volcanic ash mantle in the foreground (top left), and the ash cloud (top right). The north-eastern margin of the new crater (the horizontal line) as of September 1991, after the June 15, 1991 collapse of the old structure (bottom). Along the two incisions originating from the crater (left side of the picture) hot lahars were channeled towards the initial portion of the O'Donnel River.

Major explosive episodes are reported to have sent ashes as high as 40 km with finer particles reaching even further. Large areas of Central Plain and Zambales Provinces were at this time a scene of unprecedented destruction: a huge number of house roofs had collapsed since they were not designed to withstand the weight of several centimeters of ash blanket. The landscape and the farmland were buried under a variable thickness of ash, while human activities in commerce, agriculture and industry had come to a standstill. The drama of the homeless and jobless was the initial step of a further tragedy: farm animals were dying everywhere since food or grazing was no longer available and the agro-industrial framework of the region was entirely disrupted.

The overall volume of ejecta, including the amount deposited in Luzon and surroundings plus the finer part blown into the atmosphere, is estimated at several cubic kilometers. During site visits by helicopter huge pot-shaped valleys filled with pyroclastics were observed near the crater, with an estimated thickness of over 100 m of ashes and other ejecta.

Figure 8.7 (based on a Hazard Map of Mount Pinatubo as of July 1991, prepared by Punongbayan and Rimando of PHIVOLCS) shows isopachs of the ashfall, the location of pyroclastics and lahar (mudflow) areas around the volcano. Pyroclastic flow deposits radiate from the crater along major incisions, while isopachs of airfall ash indicate the thickness of the deposited material. Due to the westward direction of typhoon Diding in June 1991 larger quantities of ash accumulated over the Zambales side of the volcano. Finally, in black, the map shows lahar prone areas, already reached by mudflows by the end of July 1991.

After the mid-June paroxysms of the volcano, some gas and ash emission and minor seismicity continued during July through September, although at a lower intensity.

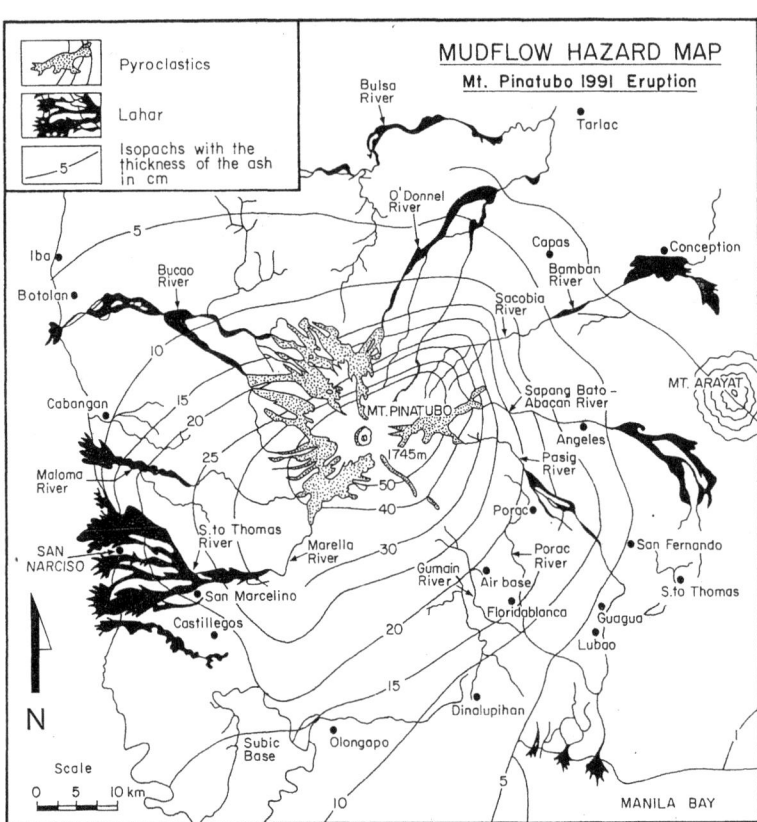

Fig. 8.7 – *Mudflow Hazard Map of Mount Pinatubo, showing isopachs of the ashfall and lahars as of July 1991 (adapted from Punongbayan and Rimando, 1991).*

8.3.2 Some considerations on the eruption and related effects

Major primary effects of the eruption were the deposition of heavy ejecta near the crater and the blanket of ashes covering an area of over 10, 000 square kilometers in Central Luzon. The entire environment sufferd badly from the eruption and disastrous consequences affected human activities, fauna and flora. Steam, sulphur dioxide, other gases and fine ash were dispersed into the atmosphere affecting the earth's climate and the ozone layer in the tropics. More dangerous and destructive for human activities and structures were lahars: mudflows of poorly-consolidated volcanic ashes mobilized by heavy rain. With a wet-concrete consistency, the slurry-like mixture began moving downslope, through minor and major streams, finally reaching the plains.

Overlapping old lahars were discovered in various parts of the flat land around the cone, as a consequence of the riverbank erosion. West of the volcano, near San Marcelino in Zambales for instance, a sequence of old mudflows was observed along Sto. Thomas River banks. The Central Plain, which is a geologic depression filled with loose Tertiary and Quaternary materials, was also shaped by the abundant products ejected by the local volcanoes and by the subsequent lahars.

By September 1991, sixty-two municipalities and two cities in the provinces of Zambales, Tarlac, Bataan and Pampanga were declared calamity areas. Of the 581 casualties 200 died as a direct effect of the eruption, 77 were victims of lahars, 211 lost their lives due to exposure to disease and malnutrition at evacuation centers and 93 died from other causes. Among the casualties in evacuation camps most were Aetas. Over one million people were affected by the eruption and its aftermath.

About 80,000 hectares of fertile land were buried under the ash blanket or lahar in the provinces of Zambales, Bataan and Pampanga (Fig. 8.2). Heavy damage was inflicted on the irrigation system, water and electricity distribution services, infrastructure in general and roads and bridges in particular. In the major industrial cities of Angeles and Olongapo most of the commercial and industrial activities were disrupted, while Clark Air Base was evacuated and later abandoned.

A number of critical lahar periods, associated with major explosive phases and tropical storms, occurred during the June-September monsoon season. According to The Manila Times (July 2, 1991), «steaming earth and boulders the size of refrigerators roared down the banks of the Porac River last Sunday (June 30), burying six Pampanga villages (Fig. 8.2) and threatening six more». On July 24, 1991, according to the same newspaper, the day before «more than 60,000 people fled their homes after massive lahar loosened by the tropical storm Herming, swept through 10 rivers in Tarlac, Pampanga and Zambales». Some 5,000 others were trapped in 12 villages in Conception (Tarlac), and Air Force helicopters were sent to rescue people who had sought refuge on roofs.

8.4 Volcanic ejecta and their mobilization

8.4.1 Pyroclastics, sediment delivery in 1991-93 and gases

The main explosions occurred June 12 through 15, 1991. A huge quantity of pyroclastics, including blocks, pumice and ashes, accompanied the numerous gray clouds which reached more than 30 km above the crater. Pyroclastic flows, channelled along narrow steep-sided valleys around the cone, soon filled river channels. Abundant ashfalls occurred during the days of the most violent explosions, blanketing the provinces of Pampanga, Zambales and Tarlac. The upstream segments of major rivers (Fig. 8.7) originating near the crater (Sacobia-Bamban, O'Donnel, Bucao, Maloma, Sto. Thomas, Gumain, Porac and Pasig), received huge quantities of pyroclastics, while topographic depressions and gullies were filled by falling eruption products and flows. By June 15, with the collapse of the crater after 16 strong explosions and the blanketing of the entire volcanic edifice and its surroundings, Mount Pinatubo entered a relaxing phase. Through July-September, the volcano cooled down while the numerous lahars activated by monsoon rains ravaged the flat land. By October 1991 the major eruption of this century in Luzon was over, but death and destruction dominated the whole environment at a distance of tens of km from the crater.

Based on data from various sources (Janda et al., 1991, Pierson et al., 1992) and on the author's considerations, pyroclastic flow deposits around Pinatubo amounted to over 7 cubic kilometers. Of this volume the quantity deposited in the catchment basins of the 5 major rivers was evaluated at 6.65 cubic kilometers (Table 8.1) and the quantity of erodible material with the potential to turn into lahars was estimated at 3.44 cubic kilometers.

The volume of sediments carried by lahars during the 1991, 1992 and 1993 rainy seasons reached 0.8, 0.55 and 0.5 cubic kilometers, respectively. The residual volume (1.86 cu. km) is expected to turn into lahars during the next years, depending on the intensity of seasonal rains. Although the revegetation and the natural compaction of ashes may partly reduce the 1994-98 mudflow hazard, the risk remains high.

TABLE 8.1 - Potential lahar sediment volumes for major Pinatubo drainages, and the sediment delivery rates from 1991 to 1993 (mcm: million of cubic meters). After *Pierson and others*, 1992

WATERSHED	Volume of pyroclastic flow deposits (mcm)	Volume of erodible PF (mcm)	Erodible pre-eruption sediments (mcm)	Potential Lahar sediment volume (mcm)	1991		1992		1993 **		Lahar deposits to date (mcm)	%	Remaining source sediments (mcm)	%
					Lahar deposits (mcm)	%	Lahar deposits (mcm)	%	Lahar deposits (mcm)	%				
O'Donnell-Bangut-Tarlac	600	240	24	264	100	38	20	8	30	11	150	57	114	43
Sacobia-Pasig-Abacan	1,600	640	64	704	210	30	110	16	100	14	420	60	284	40
Sacobia-Bamban	900	360	36	396	100		70		45		215		*	
Pasig-Potrero	500	200	20	220	50		40		55		145		*	
Abacan	200	80	8	88	60	100	0	0	0	0	60		0	
Porac-Gumain	50	50	10	60	60	100	0	0	0	0	60	100	0	0
Marella-Sto. Tomas	1,300	650	65	715	185	26	195	27	125	17	505	71	210	29
Balin-Baquero-Bucao	3,100	1,550	155	1,705	250	15	230	13	250	15	730	43	975	57
TOTAL	6,650	3,130	318	3,448	805	23	555	16	505	15	1,865	54	1,583	46

* The Sacobia-Bamban and the Pasig-Potrero Rivers shall be competing for the remaining source sediments on the Sacobia pyroclastic flow fan, with the Pasig-Potrero dominating at the moment.

** Again in 1994 destructive lahars from Pinatubo buried more than 1,000 homes killing at least 23 people. On 23 and 24 September (International Herald Tribune, September 26, 1994) 15 villages were ravaged in Porac and Bacolor districts (near San Fernando, Pampanga, Fig. 4.11) by sediments up to 4 m thick in some places. Based on information televised on September 25, about 60 thousand people fled their homes.

The overall amount of materials ejected by Pinatubo during the 1991 critical phases (including the quantities accumulated on the volcanic landscape, deposited on land and sea at greater distances from the crater and dispersed into the atmosphere) is believed to be of the order of 10 cubic kilometers.

Together with pyroclastics, steam and other gases were ejected during the eruption. As an indicator of the evolving eruption phases, the sulphur dioxide content of ash clouds was estimated (Mount Pinatubo, 1991) by means of a correlation spectrometer (COSPEC), provided by the USGS. Measurements taken from May 13, suspended for a few days during the most intensive eruption phase and resumed soon afterwards, revealed that the initial emission of 2,000 tons per day of sulphur dioxide rose to 5,000 tons per day by the end of May. The consistent increase was interpreted as a result of the magma rising towards the crater, while the sudden drop to 263 tons per day by June 5, on the other hand, was taken as an indication of the magma approaching the surface and the consequent plugging of gas passages. The monitoring of sulphur dioxide emission proved to be an excellent indicator of rising magma, thus, enabling PHIVOLCS experts to forecast the most severe explosive phase, which actually occurred during June 12 through 15.

8.4.2 Composition of the ash blanket and its effects on agriculture

According to Blong's classification of pyroclastic fragments (Blong, 1984), blocks and bombs range between 25.6 cm and 6.4 cm, lapilli between 6.4 cm and 2 mm, and volcanic ash (tephra) between 2 and 0.004 mm. Mount Pinatubo pyroclastics mostly included gravel and sand-size ejecta of porphyritic biotite-horneblende quartz latite pumice (Smithsonian Institution, 1991b). The portion of heavy products was mainly found within a 10 km radius of the crater, along major incisions and depressions. The huge quantity of fine volcanic ash was mainly deposited over the landscape around the cone and surroundings and partly injected into the atmosphere.

The presence of a considerable amount of fines explains the quantity of dust dispersed into the atmosphere and circulated around the globe, as well as the numerous, destructive lahars activated by the seasonal rains. Based on analyses by the BSWM (1991) on a few samples, the ash blanket deposited around Pinatubo's slopes mainly consisted of volcanic glass; predominantly medium to coarse silt (0.06 - 0.004 mm) and the rest of fine to coarse sand (0.06 to 2.0 mm). Some scatter of results was observed and in some locations the coarse component was greater. Deposition of the ash around the cone, during various explosive episodes, was influenced by a number of factors, namely wind direction and speed, presence and amount of gases, distance from the crater, grain size of the ash and the gradient of the natural slope.

The weekly bulletin issued by the BSWM (1991) indicated that except for nitrogen (virtually absent), the major plant nutrients P, K, Ca and Mg were adequate in the ash and in the near future should result in good crop yields with low fertilizer inputs. The presumably rich allophane content (Pinatubo Soilswatch, August 1991) of the ash was expected to cause some P fixation and slow down the mineralization of organic matter.

The report also suggested that the «most worrying aspect is the excessive sulphur and iron contents in a significant number of samples even though pH almost invariably ranges between 6 and 7.7». Another hazard from the ash was the high acidity (pH between 4.5 and 4.0) of the groundwater table at depths of about 1.0-1.5 m; thus drinking water drawn from shallow wells could cause health problems. These negative effects, according to the report, will not last long, and agriculture should benefit from the new inputs of volcanic material in the course of the next decades. The thickness of the ash cover is an essential feature in this respect; up to about 10-12 cm it can easily be mixed with the existing topsoil and later become a high fertility material.

8.5 The global effect of the eruption

The Pinatubo ash cloud, which was propelled into the atmosphere in mid-June 1991, circulated westward and was reported to have reached the Hawaiian archipelago by July 1991, after almost completely circumnavigating the globe.

Soon after the eruption scientists feared that the ash-gas cloud could produce a global cooling, large temperature variations in various parts of the world as well as a negative impact on the ozone layer.

The global cooling, which was one of the effects induced by the volcanic dust and gases in the atmosphere, is due to the partial reflection of the sun's radiation. The phenomenon, which can last a few years, is thought to have contributed to anomalous weather conditions and a temporary reduction of the greenhouse effect.

According to J. W. Waters (Earth, November 1991) ten billion tons of ash and gas were ejected by the volcano. Of this volume 20 million tons were estimated to consist of sulphur dioxide. The combination of this gas with water produces sulphuric acid droplets whose shiny surface reflects part of the incoming radiation from the sun, thus lowering average global temperatures. The quantity of sulphur dioxide spewed by Mt. Pinatubo was three times the quantity ejected by El Chichon (Mexixo, 1982), whose ash cloud volume, however, was reportedly evaluated at half a cubic kilometer, an order of magnitude smaller than that of Pinatubo. Recent deviations in global temperature are in good agreement with these observations. Cold weather is also reported to have followed the eruptions of Krakatoa in 1883 and Tambora in 1815.

Cited by Earth (November 1992), Alan Robock (University of Maryland) predicted the possibility of local warming caused by the altered wind circulation. Unusual episodes of high temperature occurred, for instance, in Italy during the winter between 1992 and 1993 and in other parts of the globe. In contrast, episodes of unexpectedly low temperatures were experienced in Europe in November 1992 and March 1993. Between 11 and 14 March 1993 the entire East coast of the United States, from Florida to Washington, was hit by unusually strong winds and abundant snow in the northernmost coastal zone, with loss of lives and several billion dollars worth of damage. New York State and Florida had blackouts and paralysis of activities for a few days. According to local authorities the last time a similar case was recorded in the same area was 105 years ago.

NASA researchers (Earth, November 1992 and July 1993) indicate that the Pinatubo eruption caused considerable interference with the atmosphere and with average global temperatures. Satellite observations from the Earth Radiation Budget Experiment indicate that aerosols from the eruption could result in the increase of the amount of solar energy reflected back into space by the upper atmosphere enough to lower the average global temperature by 0.5 degree C. This cooling is expected to affect some parts of the world more than others and to continue probably until 1994-95. It is still too early to say the last word on the effects of Pinatubo's ash-gas cloud on climate. Observations made so far are in good agreement with predictions, although the amount of data is insufficient for a firm conclusion on the influence of the volcanic eruption on climate.

Another feared consequence of Pinatubo's sulphuric acid droplets in the atmosphere is the reduction in the ozone concentration. Ozone-poor air, by increasing the quantity of ultra-violet radiation received at the earth's surface, can produce dangerous effects on the human immuno-defense system and DNA, crops and life in general.

According to J. W. Waters (Jet Propulsory Laboratory, Pasadena), low values of ozone in the Pinatubo sulphur dioxide cloud above tropical regions were detected by the Microwave Limb Sounder (MLS) installed in the NASA Upper Atmosphere Research Satellite. The observed low values are believed to be due to transport effects, possibly caused by enhanced uplift of ozone-poor air from below.

The uplift, in turn, could be associated with additional heating due to solar absorption by the aerosols produced from the Pinatubo sulphur dioxide. MLS data so far available lack the long-term record which would be needed to compare time periods before and after the June 1991 eruption (Froidevaux, Waters et al., 1994; Read et al., 1993).

Figure 8.8 (J. W. Waters, Scientific American, March 1992), shows comparative images of the globe before and after the eruption.

The orange strip (lower globe) is interpreted as the sulphur dioxide cloud, presently poor in ozone. This type of gas (violet color in the upper sketch) was abundant over tropical regions before the eruption, as emerges from a comparison of the global images.

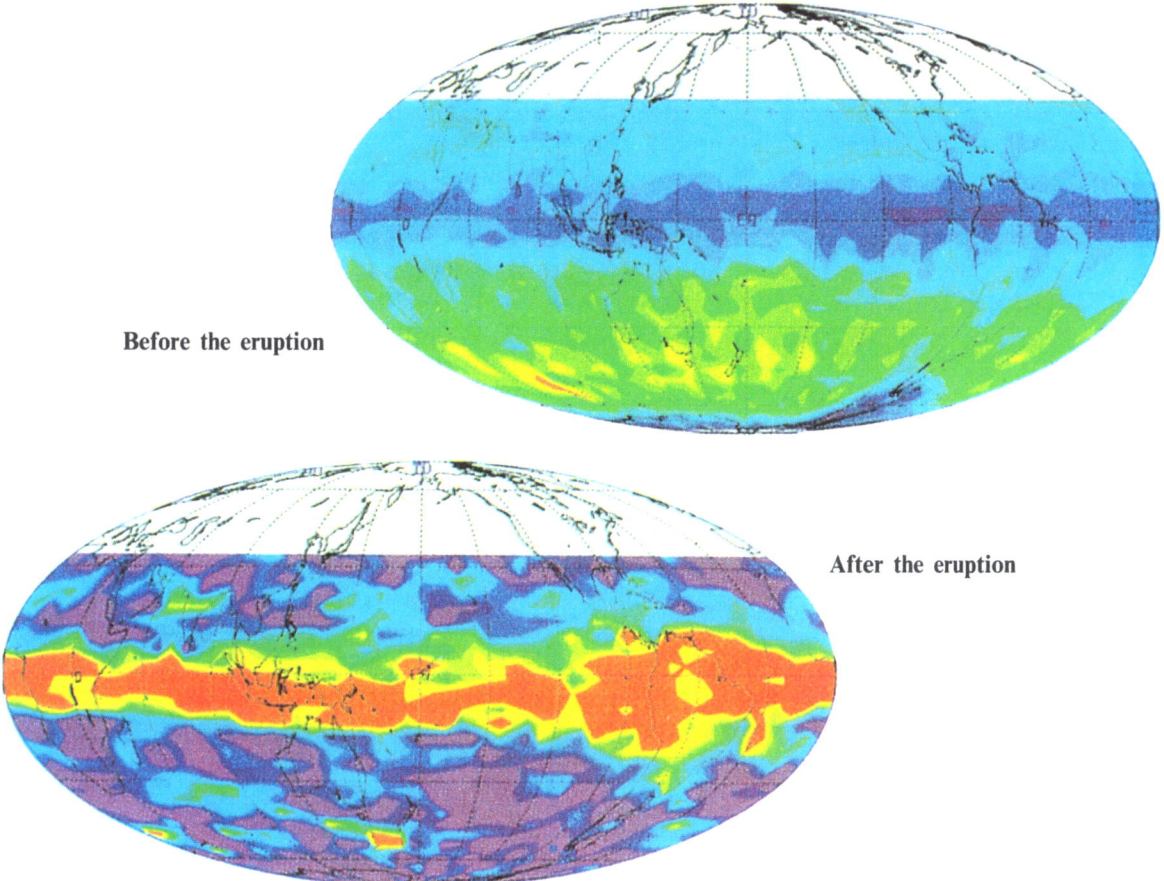

Before the eruption

After the eruption

Fig. 8.8 – *Globe images before (top) and after (bottom) the eruption of Mount Pinatubo (J.W. Waters, Scientific American, March 1992). The top picture shows in violet a continuous ozone-rich strip over the tropics. In contrast the bottom picture shows in pink an ozone-poor band caused by the enhanced uplift of ozone-poor air from below, as a consequence of additional heating due to solar absorption by the aerosols produced from the Pinatubo sulphur dioxide emission (Froidevaux L., J.W. Waters et al., 1994).*

8.6 Lahars

8.6.1 Lahar initiation and development

A description of the causative factors, composition and related flow properties of lahars is essential for understanding their initiation and development consequent to the eruption. The destructive power of the lahars and their ability to alter the geomorphology are fundamental aspects of the evolution of the landscape in Central Luzon.

Devastating hot and cold lahar around Mt. Pinatubo started during the mid-June major explosive phase, triggered either by the steam condensation or by monsoon rains; in a number of cases the two processes were associated. The whole of the June-September period was marked by numerous lahars all around the volcano and they often reached the Zambales shoreline as well as the fertile flatlands of the Central Plain.

The abundant quantity of snow-like ash deposited during the eruption had no time to become compacted under its own weight nor to be partly retained by the vegetation. Mid-June rains almost immediately filled voids in the poorly consolidated ash mantle, causing the pore-water pressure build-up and the consequent decrease of the shear strength. Of fundamental importance to the initiation of lahars and to their reaching a distance of several dozen kilometers were *a*) the loose unconsolidated condition of the ash blanket, *b*) the predominantly fine grading of the ejecta, *c*) the abundant rains brought by typhoons, *d*) the slope gradient, *e*) the presence of a well-defined natural drainage network and, *f*) a low density of vegetation in at least part of the drainage basins involved.

Further complementary aspects of mudflow evolution were the extent, geometry, shape and low roughness of the drainage network; slope gradient variations along major rivers radiating from near the cone; and shape variations of river channels with altitude and geomorphology. The loss of internal friction of the ash cover and its consequent mobilization were almost immediate, as predicted by PHIVOLCS experts, who had studied the initiation of lahars after the 1984 eruption of Mayon volcano in Legaspi (southern Luzon).

The speed of the process and the occurrence of two or more lahars per day along the same river basin were a function of the abundance and duration of rainfalls. Lahars were readily recognized by local inhabitants, due to the typical roaring sound, the gray-yellowish color, and feared because of their great destructive power.

Abundant water-borne sediments were thus deposited at distances of 20 to 45 km from the crater. Siltation begins, in fact, when slope gradient becomes minimal and a sizable reduction in the flow velocity takes place. Under these conditions the sediment load can no longer be carried and the depositional behavior prevails. This is usually associated with the widening of the riverbed and nearly level topographic conditions. When the velocity of lahars starts to decrease, water and finer suspended particles migrate upwards, while at the same time deposition of bed-load and coarser sediments begins due to the higher viscosity of the mass. A sample was taken during an on-going lahar (O'Donnel River, September 1991). After 24 hours of free sedimentation about one fourth of the volume consisted of sand and silt.

8.6.2 Hot and cold lahars and their distribution

Primary hot lahar, directly associated with the eruption, are generated during explosive phases when incandescent ash clouds carry hot rock fragments, blocks and high temperature steam. Due to the partial condensation of the steam near the surface, hot water is generated and primary hot lahars start from near the steam emission area. When this hot mixture flows during the rains or is channeled into a river, it warms up the additional cold water it gets mixed to, while the lahar progressively increases in volume. The mixed-source hot lahar can reach tens of kilometers of distance, the temperature rendering the mass less viscous and more mobile than colder mixtures.

Abundant hot lahar were described by The Manila Times on July 27, forty-five days after the most critical eruption episodes. Due to heavy rains and the fact that the Abacan River overflowed its banks, 12 villages near Mexico City (San Fernando, Pampanga) and several small villages near Angeles City were invaded by hot lahar at a distance of 25 to 40 km from the crater. The devastation induced by these hot mudflows and related flooding led to the evacuation of 11,660 people in the three provinces around the volcano, and the destruction of 7,600 houses. Small hot lahars were still occurring even at the end of September 1991 upstream of the O'Donnel River.

Cold or secondary lahars occur some time after an eruption and are triggered by heavy rains. They can include pre- and post-eruption material, reach a long distance during prolonged monsoon rains and cause enormous destruction.

Both types of lahar associated with the Pinatubo eruption had properties in common: except for a limited number of trees and smaller plants, they basically carried silt and sand, the presence of vegetation being generally confined to the period close to the explosive episodes; blocks and boulders were carried along the upper reaches of rivers, where heavy ejecta had accumulated in larger quantities; sediments deposited along the downstream portions of river channels mainly consisted of fine materials; and erosion prevailed in general along upstream sections with deposition of sediments mainly in and along the downstream river segments.

The major rivers originating from the volcano, whose beds were clogged by lahars, from northwest clockwise (Fig. 8.7) are Marella-Sto. Thomas, Balin Baquero-Bucao, O'Donnel (Bangut)-Tarlac, Sacobia-Bamban (Abacan), Pasig-Potrero and Porac-Gumain. Figure 8.9 illustrates examples of major lahars at distances ranging from a few km to about 30 km from the crater and Table 8.1 shows sediment delivery rates from 1991 to 1993.

Fig. 8.9 – *Sheet-like lahar deposits upstream of the O'Donnel River a few km from the crater (top left). The deposits are entirely composed of fine sediments; neither blocks nor debris are present. Calm lahar near Angeles City (Abacan River), with the surrounding landscape blanketed by volcanic ash (top right). Cold roaring lahar about 12-km from the crater along the O'Donnel River, near Santa Lucia (bottom left). Turbulent cold lahar at Pabanlag Bridge, near Floridablanca (bottom right). The pictures were taken in July-September 1991.*

Between June and September 1991 it was not possible to reach the upper courses of the major rivers radiating from the western and southern slopes of the crater, so the quantity of blocks and heavy ejecta carried by the lahars is poorly known. It is probable that heavy pyroclastics were deposited before reaching the mid-terminal sections of the rivers. Inspections made in September along the Zambales coastal area, where rivers are already very close to the sea, confirmed that mudflow materials consisted almost entirely of fine silt and sand. Along Sto. Thomas River, at San Marcelino (Fig. 8.7) SW of the crater, both new and old mudflows (seen in eroded riverbanks), carried small-sized pumices (3 to 10 cm in diameter) in a fine sand matrix.

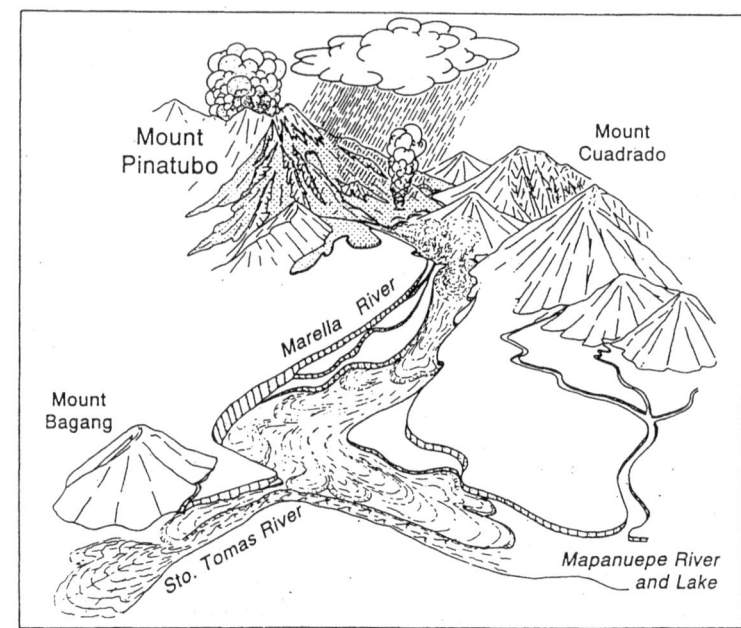

Fig. 8.10 – *Idealized sketch of Pinatubos's lahars along Marella and Sto. Thomas Rivers and Mapanuepe River damming (Umbal and Rodolfo, 1992).*

An impressive sequence of lahars along the Marella and Sto. Thomas Rivers during the 1991 rainy season was described by Umbal and Rodolfo (1992). The sizable quantity of lahars caused by the intense rainfall, due to the narrowness of the Sto. Thomas River upstream channel, was directed upstream into a tributary (Mapanuepe River) thus forming a lake (Fig. 8.10). The water volume stored by the temporary barrage later broke through the volcanic-debris dam, mobilizing the deposits of previous lahars, channeling them towards the South China Sea and heavily eroding the banks of Sto. Thomas River. This occurred during the 1991 rainy season which caused the mobilization of 185 Million cubic meters of debris along the Marella and Sto. Thomas rivers (Table 8.1) and damaged an area of 46 square kilometers.

8.6.3 Some observations on Mt. Mayon and Mt. Pinatubo lahars

The properties and behavior of mudflows generated during and after the 1984 eruption of Mayon volcano in Legaspi (southern Luzon, Fig. 2.1) were studied by PHIVOLCS researchers. Results were discussed during the First International Seminar-Workshop on Lahars and Landslides (December 1986, Legaspi) and partly summarized in the PHIVOLCS paper titled Geologic Hazard and Preparedness Systems. J.V. Umbal (1986), one of the authors, made the following observations on Mayon's mudflows: rainfall needs to be above 66 mm for lahar initiation; hot lahar have a temperature above 50 degrees C; flow is mainly laminar with boulders bobbing on top; flow velocity is between 3-5 m/sec during the initial turbulent phase; and mean grain size of particles is 0.5 mm.

There is a marked difference between the morphologies of Mt. Pinatubo and Mt. Mayon. Mount Pinatubo is larger, and the contours are quite fragmented, while its shape does not compare at all with the almost perfect conical symmetry of Mount Mayon. Within a radius of 20 km, between the crater of Mt. Pinatubo and Porac City to the east, elevations drop from 1,745 to 80 m with an average slope gradient of 8.3%. Mayon's crater, on the other hand, is at an elevation of about 2,500 m and has a continuous gradually decreasing slope to sea level about 10 km away, the average gradient being 25%. It is thus clear why the Mount Mayon lahars were able to carry heavy pyroclastics over long distances compared with the Pinatubo mudflows which were composed mainly of fine sediments in their downstream reaches. According to Umbal (1987) an area of 3.9 square km was affected by mudflows after 1984 Mayon eruption.

Lahar-invaded areas around Mount Pinatubo, by contrast, covered over 200 square kilometers by the end of the 1991 rainy season. This explains the order-of-magnitude difference in the post-eruption effects of the two volcanoes, as well as the destructive power of the Pinatubo lahars, due to the greater volume of sediments transported and deposited. Umbal estimates the volume of sediments transported by the Mayon lahars during the 1984 rainy season at 6 million cu. meters. In comparison, 805, 555 and 505 million cu. meters of sediments were carried by mudflows from Pinatubo during 1991, 1992 and 1993, respectively, and a further 1,86 cu. km is likely to be mobilized during the years 1994-1998.

8.7 Casualties and injuries

According to National Geographic magazine (December 1992), about 900 people lost their lives as a consequence of the Pinatubo explosion. Some belonged to the Aeta tribe, estimated to have numbered 12 thousand before the eruption, living in small villages within a radius of 20 km around the cone. When the situation near the crater became critical and evacuation was ordered, some Aetas refused to abandon their land; many of them died during the explosive phase, 12-15 June. Most casualties in cities and villages were caused by the collapse of roofs, houses and other buildings due to the weight of the ash. The early warning of critical eruption phases and preparedness measures formulated by PHIVOLCS, reduced the number of deaths. Thirty-nine people, however, were injured by rock fragments associated with the rain and sand which fell on June 14 near Subic Bay. According to PHIVOLCS (Mount Pinatubo, 1991), in addition to the casualties due to the direct effects of the eruption, 211 people died in evacuation camps as a result of disease and malnutrition; 94% of these were Aetas.

8.8 Damage assessment

8.8.1 General

The effects of the Pinatubo eruption included extensive damage to human activities and structures. The ash blanket, from several meters thick in the valleys near the crater down to 5 cm at an average radial distance of 40 km, affected an area of about 5,000 square kilometers. Lahars brought devastation to lands lying more than 40 km from the crater. The damage associated with the eruption was caused in three ways:

- the direct effect of volcanic ash. The weight of ash caused the collapse of roofs and often of two and three storey houses. Ash blocked roads, damaged infrastructure services, clogged natural and man-made drainage systems. Cultivated land, in particular, was badly affected by the nearly immediate destruction of crops and interruption of agricultural activities. The direct damage to agriculture concerned the flat land W and mainly E of Pinatubo (Central Plain) which used to be the richest farmland of the Philippines, because of the fertility of the soil (of volcanic origin), the abundance of water and the vicinity to Manila markets. Thus, the effect of the ash blanket on agriculture was catastrophic both in terms of immediate damage and loss of future income.
- the numerous lahars which brought devastation to lands beyond the area heavily affected by ashfalls (Fig. 8.7). During the 1991 and 1992 rainy seasons several dozen mudflows reached the flat land, villages and cities around the volcano, burying agricultural areas and destroying crops, vegetation and houses, while wreaking havoc on human activities. In the flat areas and depressions inundated by lahar near Bacolor (Fig. 8.2), groups of houses with their inhabitants were buried under 3-4 meters of sediments (The Manila Bulletin, September 12, 1991).
- lahar-related flooding of low-lying areas between 30 and 50 km from the crater. Floods are a recurrent problem in the southern part of the Central Plain (Pampanga and Bulacan provinces). The Plain is crossed by the Pampanga River (and by a complex network of tributaries) and drains into Manila Bay (Fig. 8.2). From Conception City (south of Tarlac), down to Manila there is a huge area of virtually flat land with little slope and numerous riverbeds. This drainage network badly suffered the direct impact of lahars and their delayed effects. Numerous floods occurred in this flat zone as a result of river siltation; depressions which had traditionally been used for farming or had served as small reservoirs for irrigation and as fish ponds, were inundated and locally clogged by sediments.

Fig. 8.11 — Collapsed school in Porac (top left), damaged roofs (top right) and destroyed hospital (bottom left) in Olongapo City, and collapsed roof in San Narciso, Zambales, (bottom right). Light roofs in general could not withstand the weight of the ash cover.

8.8.2 Damage to buildings and city services

Over 110,000 houses were variously damaged and 40 per cent of these were destroyed by the ash deposited on roofs (National Geographic, December 1992). Its weight caused the collapse of numerous buildings in Angeles City, 25 km East of the cone. At this distance from the crater, the ash blanket was comparatively thicker on the West side of the volcano due to typhoon winds.

The US Army Clark Air Base, 20 km SE of the cone, was evacuated first, because of the abundant ash deposits and the unbreathable air during critical eruption phases; it was later abandoned by the nearly 15 thousand soldiers stationed there and shut down probably because of the high estimated cost of cleaning and rehabilitation and the fear of future explosive episodes.

It was fortunate that the ash blanket accumulated more thickly on the western side of the volcano where cities are fewer and hence the population is comparatively smaller. Because of the tropical climate in the Philippines, roofs are merely designed to support their own weight and the loads that occur during typhoons and repair works. However, an ash blanket 10 cm thick resulted in an additional load of about 150 kg/sq.m, excessive for many light structures.

Many buildings collapsed: houses, schools, hospitals and numerous markets where roofs were merely designed to offer a protection from rain and sun. Figure 8.11 shows typical damage due to roof collapse. Water supply and sewerage systems also suffered sizable damage both from the direct impact of ashfall and for the collapse of buildings in major cities around the volcano. Telephone and power lines were also interrupted during the critical eruption phase, because dust, high temperatures and ash-covered roads prevented repair works for days.

8.8.3 Damage to roads and bridges

The road network (DPWH Report, September 1991) was directly damaged by the deposition of ash, which had to be removed like snow to enable traffic to circulate. The clogging of side drains and minor structures soon became a major problem and with the beginning of the rainy season by mid-June 1991 numerous roads were inundated.

Lahars posed a major threat to bridges by eroding around piers, by the lateral pressure on piles and by the overtopping of bridge decks. Lahars are characterized by high velocity and low viscosity upstream and by the reverse situation downstream, where riverbed gradients are minimal. Most bridges are located in areas of nearly flat topography where the speed of the flow is naturally low and the viscosity comparatively higher. When the silting process begins, before the sediment mass becomes motionless, lahars assume the consistency of a fresh and dense concrete mix, so the pressure exerted on piles can become excessive and eventually cause the collapse of the structure. Any heavy volcanic ejecta which may be carried by lahars increase the danger to man-made structures, such as houses or bridges.

Lahars are powerful landform-molding agents because of the huge quantities of sediments they carry. In the case of Bamban Bridge (Fig. 8.12) the existing structure was destroyed, the channel clogged with fine sediments and the drainage pattern of the site markedly modified. Soon after the collapse of the road and railway bridges, it was evident that new bridges would have to be longer. The geomorphological changes in other cases altered the drainage pattern by laterally moving the riverbed and hence imposing a different location for the new bridge.

Deforestation played an important role in the change of the hydrologic regime and the riverine landscape since it facilitated runoff and downslope movement of huge quantities of volcanic ash.

Damage to roads and bridges is evident from Figures 8.12-8.14, which illustrate the different types of problems caused by lahars. The sequence of figures concerns bridges located counterclockwise around the periphery of the volcanic edifice. Figure 8.12 shows the huge lahars along Bucao and Sto. Thomas rivers in Zambales (top and middle), and the damage to the road and railway bridges along the Bamban River (bottom). Figure 8.13 shows the condition of the Santa Rosa bridge (top left) in Cabangan (Zambales), about 34 km west of the Crater, where siltation clogged the river almost entirely, leaving a mere 40 cm clearance beneath the deck-beams. Maculcul bridge (top right) north of San Narciso (Zambales) suffered a similar problem. The bottom figures show the partly and totally collapsed

Fig. 8.12 – *Enormous lahars in Zambales at the Bucao River (top) where the road along the right bank was buried by sediments, and Santo Thomas River (middle) where a long concrete bridge was destroyed. Catastrophic lahars at the Bamban Bridge (Pampanga), about 10 km N of Angeles City (bottom). Both the railway and the highway bridges were washed out by the destructive power of lahars, the riverbed being clogged by about 2 m of sediments.*

Fig. 8.13 – *Damage to bridges in Zambales, between San Narciso and Botolan (see also Fig. 8.2). Clogged riverbeds by lahars in Santa Rosa (top left) and Maculcul Bridge top right), where the clearance was reduced to 40 cm and 2 m, respectively. Partly (bottom left) and totally destroyed (bottom right) Sto. Thomas bridge near San Marcelino.*

117

Fig. 8.14 – Clogged riverbed in Tarlac (top left), and partially filled riverfloor of the Bucao River (top right), on the Zambales side. Collapsed bridge span and clogged riverbed along the Superhighway around 30 km N of Manila (bottom left), and clogged floor of the Apaya River with only 1 m clearance (bottom right), north of Manila.

Fig. 8.15 – *Farmland blanketed by the volcanic ash between San Narciso and Botolan (top left), and near San Marcelino (top right), Zambales side. Partially submerged ricefield (bottom left) by grayish ash-rich waters (Zambales side), and village in Porac affected by ashfall and later on by lahars (bottom right). Due to the ash deposited during the eruption, trees look like closed umbrellas.*

bridge in San Marcelino. Figure 8.14 illustrates the condition of bridges with a significant reduction in clearance (top); two bridges North of Manila (bottom) underwent a severe reduction of the clearances, and the one on the left also suffered with the collapse of a span.

In addition to the above, other bridges and minor drainage structures were condemned; their collapse can occur any time during the next few years. The dredging of riverbeds upstream of endangered bridges, in fact, is likely to have provided a temporary extension of life only for some of them.

8.8.4 Damage to agriculture and related activities

The damage to agriculture, caused mainly by ashfalls and lahars, was estimated at over half a billion US$ (Chapter 9 gives details). The ash blanket and its partial conversion into lahars (mudflows) during the 1991 rainy season involved an overall area of 385,000 hectares, roughly bounded by Iba, Tarlac, La Paz, Mt. Arayat, Sto. Thomas, Subic Bay and San Antonio. Out of this total (ADB, 1991b) 326,000 ha were devoted to forestry, 43,000 ha to agriculture (rice, vegetables, fruit trees and sugarcane) and 16, 000 ha to fisheries. The forestry sector was spoiled mainly in the provinces of Zambales (193,200 ha) and Tarlac (83,000 ha) and 302 reforestation projects (9,595 ha) were involved.

Considerable losses occurred as a result of the destruction of agriculture facilities, irrigation systems, agri-based industrial activities and lost revenues from agricultural production.

About 16,000 ha of fishponds in Bataan, Pampanga, Tarlac and Bulacan were affected by siltation due to the ashfall and lahars. Figure 8.15 shows the effect of the ash blanket and flooding of croplands around the volcano.

8.9 Environmental impact

One major effect of the eruption was the physical impact on the environment, with widespread modification of pre-eruption landscape and destruction of the vegetation mantle. After the eruption the environment around the crater, with scattered dark- gray branchless trees, looked completely inhospitable to life. Vegetation in tropical regions recovers fast, but the recovery of rain forest which originally covered Pinatubòs landscape will take decades to centuries.

With the significant changes brought by the eruption the hydrology of the region was altered and the natural drainage system partly modified. First, a number of incised valleys near the crater were clogged with pyroclastics and heavy ashfalls. Part of this material is likely to be subject to accelerated erosion during the coming rainy seasons until a stable topography is re-established. This process of reshaping the landforms will proceed in parallel with the weathering of the newly deposited volcanic products, the growth of spontaneous vegetation and the formation of a new topsoil. Reforestation could contribute greatly and at the same time provide job opportunities for the local inhabitants.

According to the Task Force Report (ADB, 1991b) «*The DENR-Region III reported that the volcanic eruption damaged a substantial area of established forest lands in the provinces of Zambales (193,200 ha), Tarlac (83,100 ha), Bataan (33,000 ha) and Pampanga (16,800 ha) amounting to approximately 326,000 ha. This includes several DENR reforestation and social forestry projects covering about 19,402 ha in the provinces of Zambales, Tarlac, Pampanga and Bataan. A rough estimate of the cost of damage is approximately 117.87 million US$, but considering the vast devastated area and the number of years to re-establish the forest lands, the value of production loss will undoubtedly be more substantial. The DENR-PAWB also reported that perhaps the single largest unaccounted casualty in Mt. Pinatubòs recent eruption is the Philippine wildlife (flora and fauna) which are found on Mt. Pinatubo and its surrounding environs. The destruction of the endemic species in the area will surely result in the alteration of the ecological balance between flora and fauna. Endemic species in the area are the following: bats, insectivores, rodents wild boar, deer and several other species of birds, reptiles, and lesser known fauna species. Most birds and fast moving mammals have moved to safer ground, possibly northward to Tarlac, Pangasinan and Bagac-Morong area. However, the smaller species such as frogs, lizards and snakes may have been caught in the eruption. Alteration*

Fig. 8.16 – Houses surrounded by sediments in San Fernando, Pampanga (top left), and Porac (top right). Aetas people at the Evacuation Center in Porac in July 1991 (bottom).

in the hydrologic regime of rivers and streams is also inevitable due to filling up of river channels by lahar and other volcanic debris. Thermal pollution in areas directly affected by mudflow/lahar could render rivers and streams biologically dead. Streams quality monitoring is still being undertaken by the DENR to document changes in the physico-chemico-biological structure of freshwaters».

Long-term alteration occurred in the riverine habitat due to the abundant ashfalls and lahar-erosion upstream and to silting downstream. A negative influence of ash on the chemistry of river waters must have at least temporarily affected the riverine ecosystems.

Topsoil composition was altered both by the variable thickness of the ash deposits and the new minerals brought by the eruption, while shallow water tables are thought to have suffered some pollution by ash-carried chemicals.

There will be a different evolutionary trend on lands with 10-12 cm or less of ash, compared to the zones covered by heavy ashfalls. Where the cover was thin, combined effects of vegetation recovery and human agricultural activity will soon produce a higher fertility soil. Where the ash blanket is thicker a new vegetation will have to adapt to this virtually uniform, fine, nitrogen-poor material. Thus the formation of a new humus will probably take decades.

Through a comprehensive study of the potentials and limitations of the agricultural, industrial and residential areas unnecessary and uneconomic investments for social purposes can be avoided. The impact recorded on the South China Sea waters was also great. Along the Zambales shoreline fine ash from the eruption and sediments brought by lahars temporarily altered the chemistry of the waters. Corals in the area were also damaged by the turbidity induced by the finer fraction of the sediments.

Figures 8.15 and 8.16 (top) give an overview of impacts resulting from lahars on flat lands.

8.10 Social impact

There are two main social groups in the area: the Aetas predominantly living on the south-western slope of Mt. Pinatubo and the farmers mainly residing on the flatlands around the volcanic edifice. According to the DPWH (DPWH Report, 1991) nearly two million people are thought to have been affected in various ways by the eruption and its secondary effects.

An estimated 678,000 employed workers were displaced as a consequence of the eruption, thus almost doubling the number of the unemployed in the region. Primary damage due to ashfall also disrupted the activities of Clark Air Base and Subic Naval Base with immediate consequences on the purchase of local goods and the interruption of contractual services.

The 12,000 Aetas, were forced to evacuate during the main eruption episodes. According to Hiromu Shimizu (1989), who recently published an ethnographic study on the social and religious life of the Aetas, this tribe belongs to the Negrito ethnic group. Shimizu, who studied the Aetas world, sharing his life with them for a long time, reports they were forced to settle on the southwestern slopes of Pinatubo by the waves of migrants who arrived during the past century. Organized in families, small groups or temporary villages, they were used to move from place to place on the slopes of Pinatubo every five to ten years. The Aetas acquired a thorough knowledge of fauna and flora in the volcanic environment.

With their forced migration from Pinatubo and relocation in distant areas with different environmental conditions, the Aetas have suffered more heavily from the eruption than the farmers. Some of them died during the explosive phase having refused to abandon their bamboo houses. Many of them, however, gathered in evacuation camps where they suffered from malnutrition and a variety of diseases. Figure 8.16 (bottom) shows Aetas at the evacuation camp in Porac a month after the eruption.

Lowland farmers also suffered from the almost complete disintegration of their environment, as well as the destruction of their property and failure of their activities. Within a few weeks of the eruption, the natural resources on which about two million people depended were temporarily annihilated, and productive activities, agriculture and commerce as well as social life were completely disrupted.

Part of the flatland agricultural environment some distance away from the crater will recover in a few years, and farmers will resume their activities. The restoration of the Aetas environment will take much longer and the habitat will not be the same.

CHAPTER 9

THE DAMAGE TO THE ECONOMY OF THE PHILIPPINES

9.1 The trend of the economy and the impact of disasters during the period 1987-1991

According to the Asian Development Bank (ADB, 1991b), after the critical period of 1983-1986 the economy of the Philippines began to recover in 1987, slowing down again from 1989 onward. The slowdown is explained by the decrease of investments, exports and industrial growth. Beyond these factors, the progressive drop in growth rates was aggravated by a number of disasters which affected the country.

The 1989 drought lowered agricultural production with negative effects over the first half of 1990 and the consequent increase of food prices. In July 1990 a strong earthquake caused numerous casualties, widespread destruction and disastrous effects on the local economy in Central and Northern Luzon provinces. Because of these events, the growth rate of the GNP was 3.7 percent, compared to the planned 6.5 percent of the updated Medium-Term Philippine Development Plan (1988-92). The 1990-1991 Gulf crisis by increasing oil prices worsened the country's balance of payments and decreased workers' remittance from the Middle East.

The June 1991 eruption of Mount Pinatubo, however, was the most catastrophic event. First, the immediate and destructive impact of the ashfall heavily affected agriculture and private property in Central Luzon, practically destroying its economy. Second, the sequence of mudflows activated by monsoon rains brought devastation to a distance of 45-50 km from the crater, where the ashfall only had a marginal impact.

Because of the above phenomena and their combined effects, the GNP growth rate which had already lowered to 3.7% in 1990, was expected to drop to 1.1% in 1991, from the projected estimate of 1.4 (ADB, 1991b). The decelerating trend of the economy was also induced by the destruction yearly brought by typhoons, which caused several hundred Million US$ damage during each year of the period 1987-1990. Table 9.1 lists the damage to the economy of the Philippines during 1987-1991, due to these major domestic disasters and the Gulf crisis.

The loss of nearly 4 billion US$ during the period 1987-91 (Note a, Table 9.1), due to the combined effects of national disasters and the Gulf War, definitely hampered the recovery efforts of the economy; more than 3 billion of this amount occurred in 1990-91, about 1.5 billion US$ per year. Over a Gross National Product of 28.6 billion US$ in 1990, this represents a loss of more than 5 percent. This is lower than the expected 6.5 percent growth rate in 1990 (updated Medium-Term Philippine Development Plan, 1988-92), but greater than the actual 3.7 per cent growth rate in that year.

Beyond the direct damage which is clear from these figures, there were severe long-term effects due to the disruption of the human environment, which is the essential framework of an agriculturally-based economy. The earthquake and the eruption directly affected nearly 3 million people in various ways, knocking down the economy of Central Luzon, a region which accounts for 60 percent of the production of the Philippines. According to the Task Force Report, as of August 1991 (ADB, 1991b), «It should be noted that due to substantial permanent and irreparable damage (by Pinatubo's eruption) suffered by the agriculture sector in the region, the total agricultural production losses over the next five years may be valued at about 890 million US$».

TABLE 9.1 - Damage to the economy of the Philippines during the 1987-1991 period, billion US$	
Typhoons	1.26
July 1990 earthquake	0.64
Gulf crisis (1990-91)	0.38
June 1991 Mt. Pinatubo eruption	0.71
Immediate production losses due to earthquake and eruption	0.71
Total	3.70

a) typhoon-related damage in the 1991 is not included; a preliminary figure, however, puts the 1991 damage at over 250 million US$, thus bringing the total loss to nearly 4 billion;
b) Exchange rate 1 US$ = 25 Pesos (1991).

The earthquake and eruption-related production losses in the various sectors of the economy, during the period 1992-1996, are likely to exceed 2 billion US$. The total cost of the reconstruction and rehabilitation can be estimated at over five billion US$.

9.2 The damage caused by the 1990 Luzon earthquake.

Several domestic and international research groups evaluated the economic damage caused by the July 1990 earthquake. Results show a certain scatter due to different investigation methods and the composition of investigating teams. Based on these studies a comprehensive evaluation of earthquake induced damage is given below.

There was considerable damage to houses, buildings and structures in general. Baguio and the coastal area between Agoo and San Fernando in La Union suffered heavy destruction, while Cabanatuan City located near the first epicenter recorded only the collapse of the Christian School and few casualties. The liquefaction in the Central Plain, instead, involved an area of over 1,500 square km, while landslides in the mountainous provinces occurred in an area of 10,000 square km.

About 1.25 million people were affected and 100,000 houses were damaged. Most of the casualties occurred in Baguio district and several were due to landslides.

The road network was seriously affected by the collapse of bridges or their approach embankments and by the failure of innumerable slopes. Dams near the ground rupture zones and near Baguio City suffered various negative effects, such as cracks and numerous landslides within the catchment basin; a nearly complete siltation of the Ambuklao reservoir (near Baguio) forced the shutdown of electricity production and hampered irrigation activities.

The damage to the environment was severe, due to the widespread slope denudation in the Central Cordillera and Caraballo Mountains. In these two ranges, together with the sliding of the regolith layer, trees were uprooted over extensive areas. The major part of the damage to the environment was not economically quantifiable, consisting of the destruction of resources, such as topsoil and natural vegetation which are not renewable in the short term.

The overall damage, based on various sources including JICA (1990), DPWH (1991) and Author's investigation, is summarized in Table 9.2. Damage to infrastructure accounted for some 40 percent of the total, agriculture less than 10, and private property and industry sectors suffered nearly half of the total damage.

TABLE 9.2 - Damage due to the July 1990 earthquake		
		Million US$
Agriculture		57.0
Crops	22.0	
Fisheries	16.3	
Livestock	1.6	
Irrigation	4.0	
Others	13.1	
Infrastructure		273.8
Roads/highways/bridges	138.9	
Schools/hospitals/government buildings	134.9	
Private property		158.2
Industry, commerce mining and tourism		148.0
Industry and commerce	104.0	
Mining	21.1	
Tourism	22.9	
	Total	637.0

a) the total loss was rounded to 0.64 billion US$ in Table 1;
b) Exchange rate 1 US$ = 25 Pesos (1991).

9.3 The damage induced by the 1991 eruption of Pinatubo

The large ashfall which blanketed Central Luzon provinces caused primary damage followed by destructive mudflows (lahars) activated by the monsoon rains soon after the eruption. Within the heavily affected area (550,000 ha) the cover varied from several meters thick near the crater to some 5 cm at a distance of 40 km. About 385,500 ha of agricultural land (ADB, 1991b) were badly damaged by the ashfall and the later mudflows (near the downstream courses of the major rivers radiating from the crater).

Around 200,000 people were evacuated from the most critical zones near the volcano, 300 evacuation centers were established by the government and a state of calamity was declared in 58 municipalities. Agriculture was the most severely affected sector of the economy. According to the ADB, agriculture accounts for 27 percent of the GDP and over 50 percent of total employment. Half of the new jobs

TABLE 9.3 - Damage and losses due to the 1991 eruption of Mt. Pinatubo			
			Million US$
Agriculture			424.5
a) Damage to agricultural production		229.7	
1) Crops (Rice, vegetables fruit trees, sugarcane)	44.7		
2) Livestock	2.4		
3) Poultry	2.4		
4) Fishery	2.3		
5) Forestry	177.9		
b) Foregone revenue in Agricultural Production		179.6	
c) Damage to the Facilities of the Department of Agriculture		4.6	
d) Damage to Irrigation Facilities		10.6	
Industry			15.3
1) Agri-based industry	6.9		
2) Others	8.4		
Infrastructure			66.5
1) Roads, Highways and bridges	25,6		
2) Water distribution network	4.5		
3) Schools	34.8		
4) Hospitals	1.6		
Private property			205.1
		Total	711.4

Data on agriculture and industry derived from the Report of the Task Force (ADB, 1991a).
Exchange rate 1 US$ = 25 Pesos (1991).
Data expressed in million US$.

are created in the agricultural sector and most of the demand in industry and services is generated by agriculture and related activities. Agriculture accounts for 25 percent of export earnings. Four sectors of the economy were severely affected by the eruption (Table 9.3)

9.4 Overall economic losses caused by earthquake and eruption

The disasters of 1990 and 1991 were complementary, both geographically and in terms of the sectors of the economy most affected. Together they caused widespread general devastation of Central and Northwestern Luzon.

Sector	Damage by the 1990 quake	Damage by the 1991 eruption of Pinatubo	Total damage Million US$
TABLE 9.4 - Damage due to the disasters of 1990 and 1991			
Agriculture	57.0	424.5	481.5
Private Property	158.2	205.1	363.3
Infrastructure	273.8	66.5	340.3
Industry and Commerce	148.0	15.3	163.3
Total	637.0	711.4	1,348.4

Note Exchange Rate 1 US$ = 25 Pesos (1991).

Table 9.2 shows that July 1990 tremors were most destructive for the infrastructure sector (43%), and significant for the industry (23%) and private property (25%) while the damage on agriculture (9%) was marginal. The destruction associated with the quake mostly affected the mid-northern part of the Central Plain and northwestern Luzon provinces.

Table 9.3 shows that the Pinatubo eruption was catastrophic for the agricultural sector (60%), and that damage to private property was considerable (almost 29% of the total), while damage to industry and infrastructure was relatively smaller (11%). The areas most heavily affected were Pampanga and Bulacan provinces in the western part of the Central Plain and the provinces of Bataan and Zambales, south and west of the crater, respectively. Table 9.4 shows the total damage by sector.

Agriculture was most heavily affected, absorbing more than one third of the cumulative damage, followed by Private Property and Infrastructure sectors with about a quarter each. Industry and commerce suffered only about a tenth of the total damage.

CHAPTER 10

LESSONS LEARNED FROM THE 1990-1991 DISASTERS

10.1 Introduction

Natural hazard mitigation and integrated development planning are essential tools in the hands of decision makers, planners and administrators in the attempt to reduce the impact of natural disasters. In the case of the Philippines, based on the lessons learned from the 1990-91 events, the development model adopted in Central Luzon needs to be revised and alternative solutions considered. Prevention, reduction and mitigation of disaster effects (UNDRO, 1991) involve three main steps:

- Risk Assessment, which defines the types and magnitude of disasters by means of a multidisciplinary task force of geoscientists, engineers, planners, sociologists, environmentalists etc.
- Planning and Decision Making, which organizes the social response to the risks posed by disasters, after alternative strategies have been considered and the relative cost-benefit analysis evaluated;
- Implementation, which translates plans and decisions into action at various levels.

A full discussion of this complex process is beyond the scope of this book. However, the lessons learned from the earthquake and the eruption have been summarized below in three sections:

- remarks pertaining to the vulnerability of the Philippines and social response
- observations on problems related to specific aspects of Luzon's disasters
- recommendations and suggestions.

10.2 Disaster proneness of the Philippine Archipelago and social response

Some disasters which struck during the period 1987-91, such as typhoons, floodings, landslides are frequent and occur virtually every year affecting some highly vulnerable and well identified areas of the Archipelago. On the other hand, major earthquakes (which occur almost anywhere in the country) and volcanic eruptions (which mainly take place along volcanic belts), are characterized by recurrence periods of centuries in the same area, even though in the Country as a whole they recur at much shorter time intervals.

As already mentioned in Chapter 2

- 6 earthquakes of magnitudes between 7.3 and 8.3 hit the Philippines during the last 40 years, that is a strong event about every 7 years (Table 2.1);
- 41 destructive eruption episodes occurred in the Archipelago during the last 422 years (1572-1994), that is an event about every 10 years on average.

Such a national scenario, with annual and decadal recurrence of disasters, must be taken into consideration by planners and administrators. The events that affected Luzon in 1990-91 represent, of course, an unfavorable coincidence: two major disasters following one another within a year in an economically developed and densely inhabited region. It is highly improbable that such a singular sequence would recur soon somewhere else in the Country. On the other hand, earthquakes or eruptions by themselves can be even more destructive than the recent disasters in Luzon. Mount Taal (60 km south of Manila), for example, often erupted in the last few centuries, with explosions probably more powerful than that of Mt. Pinatubo and with much longer episodes of activity.

The disaster proneness of the Philippine Archipelago is a fundamental characteristic of the country, since it is located in the boundary zone of colliding plates and in the western monsoon region of the Pacific.

During the events of 1990-91, the social response in terms of mitigation measures succeeded in alleviating some of the problems. Disasters and aftermath showed, however, the limits of pre-disaster development planning and the precarious condition of people, activities and investments in various sectors. At the same time it became clear that not-allocating more resources in a regional plan for disaster prevention, reduction and mitigation can be dramatically wasteful.

Political leaders, decision-makers, administrators and the people who most suffered the effects of the 1990-91 calamities have learned the lesson and gained a better understanding of the relationship between extreme geological events and human development. Essential during 1990 and 1991 were the action of the Government in allocating resources for disaster management and the function of the National Disaster Coordinating Council (NDCC), established in 1978 to strengthen disaster control capability and preparedness.

10.3 Problems related to specific aspects of disasters

10.3.1 Landform evolution and the formation of a new regolith

G.H. Dury (Gregory, 1977) indicates that landslides and major eruptions of tephra can deliver into the natural drainage system more sediments in a single episode than the natural drainage system from an entire continent in a year. This gives an idea of the importance of the magnitude of the phenomena which affected Luzon in the 1990-91 period.

Landslide sediments mobilized by the July 1990 seismic activity in Central and Northwestern Luzon mountains were estimated at 0.4 cu. km; an additional enormous and not quantifiable volume of loosened topsoil was also produced by the quake. Pyroclastic material deposited by Mount Pinatubo was estimated at about 7 cu. km, half of which were considered erodible within a few years, depending on the intensity of rains (Chapter 8, Para. 8.4).

This huge quantity of about 4 cu. km of mobile sediments (0.4 cu. km from slides, 3.5 cu. km from tephra plus the volume of seismically loosened materials) represents a formidable hazard for flat-land areas, people and their agricultural and industrial activities. The ashfall mass, in particular, under normal weather conditions for the area, may cause significant changes in the geomorphology and hydrological regimes, and bring about river channel migrations and flooding of the lowlying lands.

The dynamics of Luzon's environment after the 1990-91 events will be strongly influenced by three interrelated phenomena: the intensity of seasonal rainfalls, the consequent sedimentation process and the formation of a new regolith layer.

The occurrence in the next few years of rainstorms with return intervals above 50-100 years could trigger another disaster by mobilizing considerable volumes of the loose sediments generated by the quake and the eruption. Relevant to the uplands and plains covered by recent ash deposits and to the scars left by the quake is the third aspect: the formation of a new regolith layer.

Slope instability phenomena in Cordillera and Caraballo Ranges (see Chapter 6) were not uniform over the landscape. Surficial soils of gently undulating areas in many cases remained as nearly undisturbed «islands», but with some local movements. Numerous steeper slopes failed, and their vegetation cover with the regolith layer slid down. Loosened soils can partly recover their particle aggregation, if not subject to the immediate increase of the erosion rate. Slide material, on the other hand, with uprooted trees and minor vegetation easily becomes a source of debris, ready for further mobilization.

This, in fact, started during the heavy rains of August through October 1990, and caused a rapid and massive downstream movement of sediments with destructive effects on infrastructure and environment. The formation of a new regolith layer over the slide areas will follow different trends depending on local conditions. In the case of scars with fresh rock outcrops, the process will take millennia; in contrast, wherever failures involved a very shallow layer or simply slope wash and the bedrock was not exposed, the formation of new regolith will be much faster.

The impact of Mount Pinatubòs ashfall was more extensive and destructive than the numerous landslides induced by the quake and affected the entire environment around the volcano. Landforms, topsoil, vegetation, wildlife, the natural drainage system, water infiltration and aquifer recharge were all heavily affected at once, and the pre-eruption environment was obliterated. The ash blanket is presently undergoing a complex and dynamic process of erosion and deposition of sediments in the flatland areas during rainy seasons. The formation of a new regolith layer is expected to be relatively fast in the areas with a thin ash cover, and will probably be completed within a decade. The new humus will be different, but in the long run it would tend to become similar to that existing before the eruption. This is because the flat land around the volcano was affected by eruptions and ensuing mudflows in the past as well. Three pathways in soil evolution are expected:

— where the ash cover is erodible and the pre-eruption topsoil will be exposed again after a few rainy seasons, or remain covered by a thin veil of ash, the formation of the new soil will take place soon;
— in areas with less than 10-12 cm ash and little erosion hazard, a natural resurgence of vegetation can be expected and the process will be speeded up in the flatlands by farmers ploughing and puddling the land;
— in zones with over 12 cm ash where normal tillage will not reach the original soil surface, a much longer period will pass before adapted vegetation types colonize the gradually weathering ash. This type of evolution is likely to follow a similar trend in areas where meters-thick ash-pockets have formed near the crater and in the flat zones on both sides of the downstream river segments, where thick layers of sediments were deposited.

These different trends of soil formation and vegetation resurgence are likely to condition the environment, agriculture and future development of Central Luzon.

Since farming activities are of primary importance for the economy of the Philippines, investments by the local Government and International Agencies are likely to continue in this sector for years to come. It is essential, therefore, that there is a portfolio of environmentally and agriculturally oriented projects with the capability of guiding human activities; this in turn should result in a proper use of available resources and safeguard natural processes and the environment. The need for multidisciplinary and multisectoral research and financing of specific projects in the area is beyond doubt.

Suitable reforestation programs could reduce the erosion rate, accelerate the stabilization of loosened soils and facilitate the formation process of a new humus-rich topsoil. Such an initiative should be associated with a watershed management project in turn accompanied by *a*) geochemical, pedological and vegetation studies, *b*) a comprehensive study on the magnitude of hydrological events during the last decades, *c*) research on evolution of the landforms (modification of river channels and migration of streambeds in the flood prone areas), and *d*) the implementation of remedial measures to decrease sediment mobility and discharge rates.

10.3.2 Agricultural development and future perspectives

The combination of the ongoing natural processes, human action through specific projects and government integrated planning could make the Central Plain a safer and productive farmland area. This complex process should achieve the following results:

— an acceptable and environmentally sound equilibrium between farmlands in the plains and the bordering mountain ranges. The progressive control of the erosion/sedimentation cycle through reforestation projects could in due course improve the safety of flood prone areas;

- the rehabilitation, modification and the reconstruction of the drainage and irrigation facilities during the next few years. This should control sedimentation and stabilize the agricultural land areas;

- the increase of the productivity of soils with the progressive weathering of the ashfall material. Based on the natural resurgence of vegetation and the ongoing researches at the BSWM and IRRI it should be possible to guide the agricultural sector towards a productive period;

- a reasonably secure distribution of land, based on the experience of the 1990-91 events. This should be achieved by facilitating resettlement in low-risk areas, through well directed investments, tax incentives and insurance programs. These would also focus the attention of the community on the importance of vulnerability reduction and compensate for damage that is likely to occur. The floodplain E and SE of the volcano will remain a dynamic system controlled by the interaction between natural phenomena (climate, volcanic activity, the evolution of Pampanga River basin) and human development. A renewed effort in terms of agrarian land reform could complete the process. This, of course, needs a strong political will and a high level commitment.

The control of inundations and sediment delivery remains essential. The selected system should be sufficiently flexible to cope with rainfall variations, different discharge conditions, sediment mobility and the monitoring action required for the safety of the population. Particular importance should be devoted to studies on the composition of the soil and its productivity, using the indications already made available by BSWM in terms of types of crop, growing conditions and use of fertilizers to speed up rehabilitation.

It is worthwhile, finally, to stress once more the importance of reforestation. Due to the indiscriminate exploitation of forest resources during the last decades, the forestry sector suffered the consequences of the eruption over an area of more than 300 thousand ha. Since reforestation is a labor-intensive activity, it should be widely adopted as a rehabilitation and disaster-prevention measure. Beyond the well known effects on the stability of slopes and the reduction of the erosion rates, it favors the recharge of aquifers, at the same time providing protection and food to wildlife. A long term return for the wood-based industry can also be expected. Last, but not least, reforestation programs can be regarded as an investment for future generations.

10.4 Recommendations and suggestions

10.4.1 Maps and data banks

The surveys of the region should be completed (physiography, geology, tectonics, hydrology, soils, vegetation, climate etc.), and multiple-hazard and landuse maps made available at municipal levels. Seismic microzoning and volcanic hazard maps are essential. A fundamental point in the vulnerability reduction process is that the only way to convince people to avoid the most hazardous areas near a volcano or an active fault, is to disseminate the proper information. People need to realize that their efforts and investment can be spoiled by natural hazards. At the present time, surveys and data banks do not cover some of the hazardous areas in Luzon.

10.4.2 Building codes and liquefaction-prone areas

Construction standards and building codes need to be revised and enforced, especially for areas where the intensity of the quake caused most damage (Baguio and Dagupan provinces). The majority of the seismically-induced damage to the infrastructure sector, in fact, can be attributed to the low construction standards and the disregard of appropriate building codes. These need to be updated on the basis of the experience of the 1990 earthquake and regulations be enforced by law and government controlling agencies.

A major and most destructive effect of the earthquake was the liquefaction in the Central Plain. Since surficial soils in this area also liquefied during previous strong earthquakes, there is a major risk that the phenomenon will occur again.

In general, land use maps could help in disseminating information on hazard zones and guiding development. Researches on former river meanders in Dagupan City (Torres et al., 1991) show that liquefaction-prone areas are often associated with river channel changes and the recent deposition of fine sediments. Based on this experience and using traditional investigation methods (drilling, SPT, etc.) the entire zone affected by liquefaction in the Central Plain should be surveyed in detail.

One method for the rapid identification of liquefaction prone areas, based on shear wave velocity, was proposed by Stokoe et al. (1988) and by Tokimatsu et al. (1991). The method is based on the correspondence of high values of the velocity of shear waves to high resistance to liquefaction and vice-versa. Thus, through the measurement of ground vibrations artificially induced a shear wave dispersion curve is determined and variations in the soil resistance to liquefaction identified.

The experience gained in the area affected by liquefaction suggests in general that compact one-two storey buildings resting on square or rectangular reinforced concrete platforms (sides ratio up to 1:2) can better stand liquefaction without suffering the severe differential settlement and tilting seen in Dagupan City after the quake. The reconstruction of damaged buildings and infrastructure cannot be confined to these simple models, however; nor can it ignore the cost of land, which is high for one to two storey buildings on a platform foundation. Driven or bored piles for some particular structures (such as bridges and high-rise buildings for example), remain the only way to carry the loads, through the liquefaction-prone layer, into a firm substratum.

10.4.3 The performance of the rigid pavement of the road network during the quake

The adoption of rigid pavements (usually made of 3 by 3.5 m cement concrete slabs cast in situ) for most roads in Central and Northern Luzon resulted in devastation by the ground shaking induced by the quake.

It is generally accepted that a well done rigid pavement can last for decades and does not need costly maintenance. Overloading by trucks, low quality of the concrete and poor construction technique, however, resulted in poor performance of these pavements even before the 1990 earthquake. The quake caused most damage a) in the Caraballo mountains where the Dalton Pass road was crossed several times by the ground rupture, b) in the sections severely affected by slides, and c) in the flatland areas where liquefaction occurred. In the liquefaction areas, several kilometers of rigid pavement had to demolished because of the longitudinal cracks and the undulations of the road surface (with about 20-25 m between crests) induced by the quake.

Flexible pavements would have suffered as well, but the damage would have been more limited, and repair works faster and cheaper. The problem, however, is complex and alternating sections of rigid and flexible pavements, depending on local conditions, could represent an acceptable solution. Flexible asphalt-concrete pavements adapt in general to various types of deformation and the extent of failure is more limited than where contiguous rigid slabs influence one another during earth shaking.

REFERENCES

ADB (1991a). *Disaster mitigation in Asia and the Pacific.* Asian Development Bank, Manila - Philippines.

ADB (1991b). *Report of the Task Force on the damage caused by the eruption of Mt. Pinatubo and proposed rehabilitation/restoration measures, August 1991.* Agriculture Department, Asian Development Bank (ADB), Manila.

ALGERMISSEN S.T., D.M. PERKINS, P.C. THENHAUS, S.L. HANSON and B.L. BENDER (1982). *Probabilistic estimates of maximum acceleration and velocity in rock in the contiguous United States. U.S.G.S.* Open-file Report 82-1033.

ALLEN C.R. (1962). *Circum-Pacific faulting in the Philippines-Taiwan region. Jour. of Geoph.* Research, 67, n. 12: 4795-4812.

BACHMAN S.B., S.D. LEWIS & W.J. SCHWELLER (1983). *Evolution of a forearc basin, Luzon Central Valley, Philippines.* Am. Ass. of Petrol. Geol., 67, n. 7: 1143 - 1162.

BARRIER E., P. HUCHON & M. AURELIO (1991). *The Philippine Fault: a key for the Philippine kinematics.* Geology, 32-35.

BESANA G.M., R.S. PUNONGBAYAN, J.V. UMBAL, J.A. DALIGDIG and B.C. BAUTISTA (1990). *Preliminary Analysis of the July 16, 1990 Luzon Earthquake Aftershock Distribution in Relation to Ground Rupture.* GEOCON '90. The Third Annual Geological Convention, UP-NIGS Quezon City, December 1990.

BERANZOLI L., D. GIARDINI and N.A. PINO (1993). *Seismogram Processing at MedNet. Computers & Geosciences*, Vol. 19, No. 2.

BLONG R.J. (1984). *Volcanic Hazards. A Sourcebook on the Effects of Eruptions.* Academic Press, Sydney.

BONATTI E. (1994). *The Earth's Mantle below the Oceans.* Scientific American, March 1994.

BSWM (1991). *Pinatubo Soilswatch* (August 9, 1991). Vol 1 No. 3,. A weekly bulletin of the Bureau of Soils and Water Management. Quezon City, Philippines.

CORNELL C.A. (1968). *Engineering seismic risk analysis.* Bull. Seis. Soc. Am. 58, 1583-1606.

CROSBY A W. (1992). *The Columbian Exchange. Biological and Cultural Consequences of 1492.* (Italian translation by Einaudi, Torino, 1992).

CROZIER M.J. (1986). *Landslides. Causes, Consequences & Environment.* Edited by M.J. Crozier, New Zealand.

DPWH (1991). *Assessment of damage to infrastructure caused by the June 1991 eruption of Mount Pinatubo (September 1991).* Department of Public Works & Highways, Philippines (unpublished).

EARTH (November 1992). *Global Cooling by Tom Waters.*

EARTHQUAKE and TSUNAMI (1990). *Department of Science and Technology* (PHIVOLCS), Manila, Philippines.

FAO (1991). *Report of the mission for the formulation of large-scale rehabilitation projects following volcanic eruption.* (TCP/PHI/0155-E-). FAO, Rome, October 1991 (unpublished).

FROIDEVAUX L., J.W. WATERS, W.G. READ, L.S. ELSON, D.A. FLOWERS and R.F.JARNOT (1994). *Global ozone observations from UARS MLS: an overview of zonal mean results. Jet Propulsory Laboratory, California Institute of Technology, Pasadena, California. Submitted to the Journal of the Atmospheric Sciences Special Issue on the Upper Atmosphere Research Satellite* (March 1994).

GEARY E.E. and R.K. KAY (1983). *Petrological and geochemical documentation of ocean floor metamorphism in the Zambales Ophiolite, Philippines. The Tectonic and Geologic Evolution of Southeast Asian Seas and Islands (Part 2).* Geophysical Monograph 27. American Geophysical Union, Washington, D. C.

GIARDINI D. (1991). *MedNet: Network Configuration. I Workshop on MedNet: The Broad-band Seismic Network for the Mediterranean.* September 1990. OCSEM Erice (Sicily), Italy.

GREGORY K.J. (1977). *River Channel Changes.* John Wiley & Sons.

HAMBURGER M.V., R.K. KARDWELL and B.L. ISACKS (1983). *Seismotectonics of the Northern Philippine Island Arc. The Tectonic and Geologic Evolution of Southeast Asian Seas and Islands (Part 2).* Geophysical Monograph 27. American Geophysical Union, Washington, D. C.

HARGROVE T.R. (1991). *The Mysteries of Taal.* Bookmark Publishing, Manila.

HARP E.L., R.C. WILSON and G.F. WIECZOREK (1981). *Landslides from the February 4, 1976 Guatemala Earthquake.* U.S.G.S. Professional Paper 1204-A.

IMAI H., R.S. GATUS, M.B. COLLADO (1991). *Disturbance of layer stratification in soil profile caused by liquefaction (observation in Luzon Island, Philippines, after the 1990 earthquake).* BSWM, Dept. of Agriculture, Elliptical Road Corner, Visaya Ave., Diliman, Quezon City, The Philippines.

JANDA R.J., J. MAJOR, K.M. SCOTT, G. BESANA, J.A. DALIGDIG and A.S. DAAG (1991). *Lahars accompanying the mid-June 1991 eruptions of Mount Pinatubo, Tarlac and Pampanga provinces, the Philippines (abstract)*, EOS, Transactions of the American Geophysical Union, vol 72, No. 44 p. 62-63.

JICA (1990). *Report of Japan Disaster Relief Team (JDR) on the Earthquake in Philippines of July 16, 1990.* August 1990, by Japan International Cooperation Agency (JICA); unpublished.

LEWIS S.D. and D.E. HAYES (1983). *The Tectonics of the northward propagating subduction along Eastern Luzon, Philippine Islands. The Tectonic and Geologic Evolution of Southeast Asian Seas and Islands (Part 2).* Geophysical Monograph 27. American Geophysical Union, Washington, D. C.

MORELLI A. (1991). *MedNet: Data Management. I Workshop on MedNet: The Broad-band Seismic Network for the Mediterranean*, September 1990. OCSEM Erice (Sicily), Italy.

MOUNT PINATUBO (1991). *Mount Pinatubo Wakes from 600 Years Slumber. Brochure produced by PHIVOLCS and financially supported by the Asia Foundation.* 149 (2), 153-164.

NATIONAL GEOGRAPHIC (May 1992). *The great eclipse (A Volcano's Bad Timing)*, by R.H. RESSMEYER, Vol 181, No. 5, May 1992.

NATIONAL GEOGRAPHIC (Dec. 1992). *Crucibles of creation - VOLCANOES*, by N. GROVE. Vol. 182, No.6, December 1992.

OBSERVER (PHIVOLCS), 1990. *Liquefaction in Panay Island. Quarterly Newsletter of the Philippine Institute of Volcanology and Seismology*, April-June 1990.

PHIVOLCS (1987). *Geologic hazards and disaster preparedness systems.* Department of Science and Technology, Philippine Institute of Volcanology and Seismology.

PIERSON T.C., R.J. JANDA, J.V. UMBAL, and A.A. DAAG (1992). *Immediate and long-term hazards from lahars and excess sedimentation in rivers draining mount Pinatubo, Philippines.* United States Geological Survey. Water Resources Investigation Report 92-4039, 37 p.

PINET N. and J.F. STEPHAN (1989). *Polyphased tectonic history in a wrench faulting region: example of the Philippine Fault in Ilocos Foothills* (Luzon Philippines). 28th International Congress, Washington D.C.

PINET N. and J.F. STEPHAN (1989). *The Philippine Wrench Fault System in the Ilocos Foothills, Northwestern Luzon, Philippines.* Institute de Geodynamique, URA-CNRS, Universite' de Nice - Sophia Antipolis, Nice, France.

PUNONGBAYAN R.S., R.E. RINALDO, J.A. DALIGDIG, G.M. BESANA, A.S. DAAG, T. NAKATA, T. HIROYUKI (1991). *Ground Rupture of the July 1990 Luzon Earthquake. GEOCON '90.* The Third Annual Geolog. Convent., UP-NIGS Quezon City, 1990.

PUNONGBAYAN R.S., R.C. TORRES (1990). *Correlation of River Channel Reclamation and Liquefaction Damage of the 16 July 1990 Luzon Earthquake in Dagupan City, Philippines.* GEOCON '90. The Third Annual Geological Convention, UP-NIGS Quezon City, December 1990.

PUNONGBAYAN R.S. and J.V. UMBAL (1990). *Overview and impacts of the July 16, 1990 Luzon Earthquake.* GEOCON '90. The Third Annual Geological Convention, UP-NIGS Quezon City, December 1990.

READ W.G., L. FROIDEVAUX and J.W. WATERS (1993). *Microwave Limb Sounder Measurement of Stratospheric Sulphur Dioxide from the Mt. Pinatubo Volcano.* Geophysical Research Letters, vol. 20, n. 12, pp. 1299-1302.

REPETTI W.C. (1946). *Catalogue of Philippine Earthquakes, 1589-1899.* Bull. Seismol. Soc. Am. V. 36. pp. 133-322.

RINGENBACH J.C., N. PINET, J. MUYCO & E. BILLEDO (1991). *Surface rupture associated with the July 16, 1990 Earthquake on the Philippine Fault, Central Luzon, Philippines.* C. R. Acad. Sci. Paris, t 312, Serie II.

RINGENBACH J.C., N. PINET, J. DELTEIL, J.F. STEPHAN (1992). *Analyse des structures engendrees en regime decrochant par le seisme de Nueva Ecija du Juillet 1990, Luzon, Philippines.* Bull. Soc. Geol. France. 1992 t 163, n. 2.

SCIENTIFIC AMERICAN (March 1992). *Volcanic Disruption* by John Horgan.

SEASEE (1985). *Catalogue of the Philippine Earthquakes 1589- 1983.* (Southeast Asia Association of Seismology and Earthquake Engineering), June 1985.

SEED H.B., I.M. IDRISS, (1967). *Analysis of Soil Liquefaction: Niigata Earthquake.* J. SMFD. ASCE Vol. 93, SM 3 (1967).

SHARP R.V. *Comparison of 1979 Surface Faulting with Earlier Displacements in the Imperial Valley.* The Imperial Valley, California, Earthquake of October 15, 1979. USGS Professional Paper 1254.

SHARP R.V. (1990). *Surface Faulting. The July 16,* 1990, Luzon Earthquake, USGS & PHIVOLCS (Final Report), October 1990.

SHIMIZU H. (1989). *Pinatubo Aytas: Continuity and change.* Ateneo de Manila University Press, Quezon City.

SMITHSONIAN INSTITUTION (1991b). *Bulletin of the Global Volcanism Network,* Vol. 16 N. 7.

STOKOE K.H. II, J.M. ROESSET, J.G. BIERSCHWALE and M. AOUAD (1988). *Liquefaction potential of sands from shear wave velocity.* Proceedings 9 WCEE, vol. 3, pp. 213-218.

Stop Disasters (IDNDR), N. 13 May-June 1993.

Stop Disasters (IDNDR), N. 15 September-October 1993.

Stop Disasters (IDNDR), N. 17 January-February 1994.

SU S.S. (1988). *Seismic hazard analysis for The Philippines,* Natural Hazards 1, by Kluwer Academic Publishers.

TAYLOR B. and D.E. HAYES (1983). *Origin and history of the South China Sea Basin. The Tectonic and Geologic Evolution of Southeast Asian Seas and Islands (Part 2).* Geophysical Monograph 27. American Geophysical Union, Washington, D. C.

TORRES, R.C., R.S. PUNONGBAYAN, K.S. RODOLFO, R.A. ALONSO, L.O. PALADIO, J.V. UMBAL (1991). *Fluvial Meander Sedimentation as a Locus Determinant for Liquefaction Damage in Dagupan City,* Pangasinan. GEOCON '90, Geological Society of the Philippines.

TOKIMATSU K., S. TAMURA & S. KAWAYAMA (1991). *Liquefaction Potential Evaluation Based on Rayleigh Wave Investigation and Its Comparison With Field Behavior. Proc. Sec. Int. Conference on Recent Advances in Geotechnical Earthquake Eng. and Soil Dynamics.* March 1991, St Louis Missouri, Paper N. 3.8.

TOKIMATSU K., S. MIDORIKAWA and S. TAMURA (1991). *Preliminary Report on the Geotechnical Aspects of the Philippine Earthquake of July 16, 1990. Proceedings, Second International Conference on Recent Advances in Geotechnical Earthquake Engineering and Soil Dynamics.* March 11-15, 1991, St. Louis, Missouri, Paper N. LP26.

TOKIMATSU K., and Y. YOSHIMI (1984). *Criteria of Soil Liquefaction with SPT and Fines Content,* VII WCEE, San Francisco, 1984.

UMBAL J.V. and K.S. RODOLFO (1992). *The 1991 Lahars of Southwestern Mount Pinatubo, Philippines and the Evolution of Lahar-dammed Mapanuepe Lake.*

UMBAL J.V., (1986). *Proceedings of the First Intern. Seminar-Workshop on Lahars and Landslides.* December 1986. Legaspi City, Albay, Philippines (unpublished).

UMBAL J.V., (1987). *Recent Lahars of Mayon Volcano. Geologic Hazards and Preparedness Systems,* PHIVOLCS.

UN (1993). *The Global Partnership for Environment and Development.* A guide to Agenda 21. United Nations, New York 1993.

UN/IDNDR (1991). *Overall Programmes for Disaster Reduction in the '90s (1990-2000).* IDNDR Secretariat, United Nations, Geneva Switzerland.

UNDRO (1991). *Mitigating Natural Disasters. Phenomena, Effects and Options.* A Manual for Policy Makers and Planners.

USGS-PHIVOLCS, 1990. *The July 16, 1990, Luzon Earthquake. Final Report,* October 1990 (Unpublished).

U.S. DEPARTMENT OF THE INTERIOR, Geological Survey. *NEIC Quick Epicenter Determination* N. 0-204, July 23, 1990. Worldwide location of epicenters from two hours before to 16 hours after the July 16, 1990 Earthquake in Luzon.

VARNES D.J., (1978). *Slope movement types and processes,* in SCHUSTER R.L. and KRIZEK R.J. eds., *Landslide Analysis and Control, Transportation Research Board Special Publication 176.* National Academy of Sciences.

WATERS J.W., L. FROIDEVAUX, W.G. READ, G.L. MANNEY, L.S. ELSON, D.A. FLOWER, R.F. JARNOT & R.S. HARWOOD (1993). *Stratospheric ClO and Ozone from the Microwave Limb Sounder on the Upper Atmosphere Research Satellite.* NATURE, International Weekly Journal of Science, vol. 362 n. 6421, 15 April 1993.

WIECZOREK G.F., R. ARBOLEDA and B. TUBIANOSA (1990). *Liquefaction and landsliding from the July 16, 1990, Luzon, Philippines Earthquake.* GEOCON '90. The Third Annual Geological Convention, UP-NIGS Quezon City, December 1990.

WIELANDT E. (1991). *Very-Broad-Band Seismometry. I Workshop on MedNet: The Broad-band Seismic Network for the Mediterranean,* September 1990. OCSEM Erice (Sicily), Italy.

WORKSHOP ON MEDNET, 1990. *The broad-band seismic network for the Mediterranean*, September 1990. Sponsored by ICSC and ING, Ed. Il Cigno Galileo Galilei, CCSEM Erice (Italy).

WORLD BANK, 1990. *Managing Natural Disasters and the Environment*, (edited by A. Kreimer and M. Munasinghe. Published by The Environmental Policy and Research Division, The World Bank.

WORLD DEVELOPMENT REPORT, 1992. *Development and the environment.* Published for the World Bank, Oxford University Press.

WORLD MAP OF NATURAL HAZARDS. *Munich Reinsurance Company*, Munich 1988.

WOLFE J.A and S. SELF (1983). *Structural Lineaments and Neogene Volcanism in Southwestern Luzon. The Tectonic and Geologic Evolution of Southeast Asian Seas and Islands.* Part 2. Geophysical Monograph 27, American Geophysical Union.

SUGGESTED FURTHER READINGS

Chapter 1

1992 IPCC SUPPLEMENT - *Scientific Assessment of Climate Change.* Intergovernmental Panel on Climate Change. WMO-UNEP.

CLIMATE CHANGE: *Science, Impacts and Policy. Proceedings of The Second World Climate Conference.* Edited by J. JAGER and FERGUSON H.L, 1991. *Cambridge University Press.*

EARTH, September 1993, *Earth Beat - Past Climates Hold Clues to Global Warming.*

Chapter 2

TAIAG J.C. and R.A. RINALDO *Assessment of Public Awareness Regarding Geologic Hazards* (1987). Geologic Hazards and Preparedness Systems (PHIVOLCS).

EARTH, July 1993. *The Deadly Swept Power of Away Tsunamis*, by DVORAK J. and PEEK T.

Chapter 3

YU S.B., D.D. JACKSON, C.C. LIU & G.J. YU (1988). *Dislocation Model for a Crustal Deformation in the Longitudinal Valley Area, Eastern Taiwan - Inter. Symp. Geodynamic Evolution of Eastern Eurasian Margin.* Paris.

BARCELONA B.M. (1986). *The Philippine Fault and its tectonic significance.* Mem. Geol. Soc. China, n. 7.

AURELIO M.A., C. RANGIN, E. BARRIER and C. MULLER (1990). *Tectonics of the Central Segment of the Philippine Fault: a Young Strike-Slip Fault.* C.R. Acad. Sci. Paris, t 310, Serie II, p 403-410.

Chapter 4

NEWHALL C. (1990). *Differential Ground Shaking During the July 16, 1990 Luzon Earthquake.* GEOCON '90. The Third Annual Geological Convention, UP-NIGS Quezon City, December 1990.

THE KILLER QUAKES, 16 July 1990. *Dept. of Science and Technology*, PHIVOLCS.

NAKATA T., H. TSUTSUMI, R.S. PUNONGBAYAN, R.E RIMANDO, J.A. DALIGDIG and A. DAAG, 1990. *Surface Faulting Associated with The Philippine Earthquake of 1990*, Jour. Geogr., V 99:5

AURELIO M.A., C. RANGIN, E. BARRIER and M. MULLER (1990). *Tectonics of the Central Segment of the Philippine Fault: a Young Strike-Slip Fault*. C.R. Acad. Sci. Paris, t 310, Serie II, p 403-410.

YU S.B., D.D. JACKSON, CC. LIU & G.K. YU (1988). *Dislocation Model for a Crustal Deformation in the Longitudinal Valley Area, Eastern Taiwan - Inter. Symp. Geodynamic Evolution of Eastern Eurasian Margin*. Paris.

NATIONAL GEOGRAPHIC. EARTHQUAKE. *Prelude to the Big One?* (October 17, 1989, Loma Prieta Earthquake). Vol.177, N. 5, May 1990.

DRAKOPOULOS J.K. and G.N. STAVRAKAKIS (1991). *Rupture Complexity and Synthesized Strong Ground Motion. Earthquake Hazard Assessment*. Ed. R. Fantechi and M.E. Almeida- Teixeira, Commission of the European Community.

KANAMORI H. (1981). *The Nature of Seismic Patterns Before Major Earthquakes. In Earthquake Prediction*. Maurice Ewing Ser. 4, Am. Geophys. Union.

KANAMORI H. and G.S. STEWART (1978). *Seismological Aspects of the Guatemala Earthquake of February 4*, 1976, J. Geophys. Res., 83.

Chapter 5

TORRES R.C., A.S. PUNONGBAYAN, K.S. RODOLFO, R.A. ALONSO, L.O. PALADIO and J.V. UMBAL (1990). *Fluvial Meander Sedimentation as a Locus Determinant for Liquefaction Damage in Dagupan City, Pangasinan. GEOCON '90*. The Third Annual Geological Convention, UP-NIGS Quezon City, December 199

TOKIMATSU K. and Y. YOSHIMI (1981). *Field Correlation of Soil Liquefaction with SPT and Grain Size*. Ist Inc. Conf. on Rec. Adv. in Geot. Earth. Eng. & Soil Dyn., St. Louis, Missouri.

TOKIMATSU K. and Y. YOSHIMI (1983). *Empirical Correlation of Soil Liquefaction Based on SPT N-value and Fines Content*. SF, Vol. 23 N. 4, 1983.

SEED H.B. and I.M. IDRISS *Evaluation of Liquefaction Potential of Sand Deposits Earthquakes. Symp. on Liquef. Probl. in Geot. Engin*. ASCE, Philadelphia (1976a).

SEED H.B. and I.M. IDRISS *Evaluation of Liquefaction Potential of Sand Deposits Based on Observations of Performance in Previous Earthquakes*. ASCE Convent. and Exposit., St. Louis, Missouri (1981).

SEED H.B. and I.M. IDRISS *Ground Motions and Soil Liquefaction During Earthquakes*. EERI, Berkeley, California (1982).

SEED H.B., I.M. IDRISS, I. ARANGO *Evaluation of Liquefaction Potential Using Filed Performance Data*. J. GED., ASCE Vol. 109, GT 3 (1983).

TOKIMATSU K. (1991). *Liquefaction potential Evaluation based on Rayleigh Wave Investigation and Its Comparison With Field Behavior*. Proc. Sec. Int. Conf. on Recent Advances in Geotech. Earthq. Eng. and Soil Dynamics. March 1991, St. Louis, Missouri, Paper N. 38.

TOKIMATSU K. (1991). *Considerations to Damage Patterns in the Marina District During the Loma Prieta Earthquake Based on Rayleigh Wave Investigation*. March 1991, St. Louis, Missouri, Paper N. LP 13.

PUNONGBAYAN R.S. and J.V. UMBAL (1990). *Overview and Impacts of the July 16, 1990 Luzon Earthquake. GEOCON '90*. The Third Annual Geological Convention, UP-NIGS Quezon City, December 1990.

Chapter 6

REYES S.F. (1990). *Estimating Peak Accelerations from Landslides Caused by the July 16, 1990 Luzon Earthquake*. GEOCON '90. The Third Annual Geological Convention, UP-NIGS Quezon City, December 1990.

ARBOLEDA R. (1990). *Slope Failure during the July 16, 1990 Earthquake. GEOCON '90. The Third Annual Geological Convention*, UP-NIGS Quezon City, December 1990.

GERVASIO F.C., L.L. MORALES. *Regional Geology Studies for Slope Protection of Kennon Road*. Gervasio F.C. and Associates, Manila, January 1991.

LUNDGREN LILL (1978). *Studies of Soil and Vegetation Development of Fresh landslides scars in the Mgeta Valley, Western Ulunguru Mountains, Tanzania*. Geografiska Annaler - 60 A (1978) 3-4.

ALGERMISSEN S.T. (1984). *Integration, Analysis and Evaluation of Hazard Data. Proceedings of the Geologic and Hydrologic Hazard Training Program*. Open File Report 84-760. USGS, Reston, Virginia.

THENHAUS P.C. (1984). *Geologic Evidence of Earthquakes: Some Geologic Studies and Applications in Estimating the Seismic Hazard. Proceedings of the Geologic and Hydrologic Hazard Training Program*. Open File Report 84-760. USGS, Reston, Virginia.

KEEFER D.F. (1984). *Landslides caused by Earthquakes. Proceedings of the Geologic and Hydrologic Hazard Training Program*. Open File Report 84-760. USGS, Reston, Virginia.

Chapter 7

HIRANO S., T. NAKATA, A. SAGAWA (1986). *Fault Topography and Quaternary Faulting Along the Philippine Fault Zone, Central Luzon, The Philippines*. Journ. Geography V. 95.

Chapter 8

EARTH. *Fountains of Fire*, by M. EUSTIS. March 1993.

MOUNT PINATUBO ERUPTION 1991: *Chronology of Events*. AGID News (Official Newsletter of the Association of Geoscientists for International Development. N 67/68, August/November 1991.

DE BOER J., L.A. ODOM, P.C. RAGLAND, F.G. SNIDER and N.R. TILFORD (1980). *The Bataan Orogen, Eastward Subduction, Tectonic Rotation and Volcanism in the Western Pacific (Philippines)*, Technophysics, Vol. 67, N. 3-3.

HOBLITT R.P., E.W. WOLFE, A.B. LOKHART, G.E. EVERT, T.L. MURRAY, D.H. HARLOW, J. MORI, A.S. DAAG, TUBIANOSA B.S. (1991). *1991 Eruptive behavior of Mount Pinatubo (Philippines)*. EOS Transactions of the American Geophysical Union, Vol. 72, N. 44.

PINATUBO VOLCANO OBSERVATORY TEAM, 1991. *Lesson from a Major Eruption: Mount Pinatubo, Philippines*. EOS Transactions of the American geophysical Union, Vol. 72 N. 49.

RODOLFO K.S. and J.V. UMBAL (1992). *Catastrophic lahars on the Western Flanks of Mount Pinatubo (Philippines)*. Proceedings of the Workshop on the Effects of global Climate Change on Hydrology and Water Resources at a Catchment Scale, February 1992, Tsukuba Science City, Japan.

SCOTT K.M, R.P. HOBLIT, J.A. DALIGDIG, G. BESANA and B. TUBIANOSA (1991). *15 June 1991 Pyroclastic Deposits at Mount Pinatubo (Philippines)*. EOS Transactions of the American geophysical Union, Vol 72, N. 44.

LIST OF FIGURES

LIST OF TABLES

ACRONYMS AND ABBREVIATIONS

ADB	Asian Development Bank
AQU	L'Aquila, Italy (MedNet Station)
CARP	Comprehensive Agrarian Reform Program
COSPEC	Correlation Spectrometer
BNI	Bardonecchia, Italy (MedNet Station)
BSWM	Bureau of Soils and Water Management
DENR	Department of Environment and Natural Resources
DHA	Department of Humanitarian Affairs
DPWH	Department of Public Works and Highways
ESCAP	Economic and Social Commission for Asia and Pacific
FAO	Food and Agriculture Organization
GNP	Gross National Product
GDP	Gross Domestic Product
IBRD	International Bank for Reconstruction and Development
IDNDR	International Decade for Natural Disasters Reduction
ING	Istituto Nazionale di Geofisica (Rome, Italy)
JICA	Japan International Cooperation Agency
MedNet	Mediterranean Network
NASA	National Aeronautics and Space Administration
NEDA	National Economic and Development Authority
NGO	Non-Government Organization
NEIC	National Earthquake Information Center
PAGASA	Philippine Atmospheric, Geophysical, Astronomical Services Administration
PAWB	Parks and Wildlife Bureau
PHIVOLCS	Philippine Institute of Volcanology and Seismology
SAR	Search and Rescue
SEASEE	Southeast Asia Association of Seismology and Earthquake Engineering
SPT	Standard Penetration Test
UNDP	United Nations Development Programme
UNDRO	United Nations Disasters Release Operations
VSL	Villasalto Italy, (MedNet Station)
USGS	United States Geological Survey
WHO	World Health Organization
WMO	World Meteorological Organization

GLOSSARY

Allophane An amorphous hydrated allumino-silicate mineral present in clay soils.

Andesite A fine-grained effusive rock mainly made of andesine and mafic minerals.

Astenosphere The plastic layer of the earth immediately beneath the lithosphere.

Astrobleme An ancient impact crater from the collision of a meteorite on earth.

Basalt An effusive dark rock in a glassy or fine-grained mass, mainly composed of calcic plagioclase and pyroxene, with or without olivine.

Basement rock Igneous or metamorphic rock complex, usually unconformably overlain by sedimentary rocks.

Batholith A large plutonic rock mass intruded as a result of the fusion of older rocks and mainly composed of granodioritic and quartzo-monzonitic minerals.

Benioff zone The inclined plane descending beneath the continent (as a result of the subduction process) where earthquake foci cluster.

Calc-alkaline Igneous rocks with the weight percentage of silica between 56 and 61 in the presence of an equal percentage by weight of Calcium monoxide and Potassium plus Sodium monoxides.

Conglomerate A sedimentary rock made of cemented, coarse-grained, rounded and gravel to pebble-sized rock fragments.

Cost of the damage The overall cost of the damage induced by a disaster. It includes the cost of the damage generally affecting the human environment (houses, infrastructure and working activities) plus the cost which can be attributed to the destruction of the physical and biological environments. The latter costs (due to various types of environmental impact) are more difficult to evaluate and, thus, often inaccurate. See also economic losses.

Crustal plates Blocks 60-100 km thick forming the earth's crust or lithosphere and usually named macro or micro plates based on their size.

Cyclic ground shaking The motion associated with earthquakes and consisting of cycles of displacement of alternating direction.

Cyclone An atmospheric violent phenomenon due low-pressure conditions during which the air rapidly moves in a circular direction.

Dacite An effusive, fine-grained, rock made of plagioclase, quartz and piroxene (it generally corresponds to the composition of andesite in the domain of intrusive rocks).

Differential settlement The settlement of a foundation with different magnitude between various parts of the structure. The differential settlement can create greater distressing effects on the superstructure of a foundation than the total settlement.

Dolerite An intrusive rock with a composition similar to that of basalts.

Earthquake generator or source zone A geologic structure along which earthquakes can be generated.

Economic losses Losses due to natural disasters and generally affecting man, man-made structures and human activities. Economic losses do not include the damage to the environment, as for example the destruction of a forest and the life in it during a volcanic eruption. See also the cost of damage.

Effusive rocks Igneous rocks deriving from the cooling of lavas ejected at the surface of the earth.

Environment The complex of conditions (physical, chemical, biological, social, cultural, political and economical) that characterize a certain area and influence life and development of local inhabitants.

Epicenter The point of the surface of the earth directly above the focus of an earthquake.

Fault A fracture of the earth's crust along which relative displacement has taken place.

Flexible pavement A road pavement made of flexible layers (natural granular materials or crushed rock fragments either untreated or bound with asphalt). The definition is used in contrast with rigid pavements, which are made of reinforced concrete slabs.

Focal mechanism The mechanism describing the block's motion generating an earthquake.

Focus (or hypocenter) The point from which seismic waves are generated during an earthquake. Directly above the focus on the ground surface the epicenter is located.

Granite An intrusive, coarse-grained, plutonic rock mainly made of quartz, alkali feldspar and sodic plagioclase. Granitic rocks are the major constituents of batholiths which are huge plutonic bodies resulting from the cooling of magmas intruded into pre-existing rocks.

Granodiorite An intrusive, coarse-grained, rock mainly made of quartz, plagioclase and potassium feldspar.

Ground rupture The rupture of the ground surface associated with an earthquake.

Habitat An area and its local conditions with the capability to sustain life needs of a biological population.

Hypocenter See focus.

Hurricane See tropical cyclone.

Ignimbrite An effusive rock associated to the deposition of volcanic ash, lava and dense clouds of high-temperature volcanic glass.

Internal friction The mechanical resistance of contiguous soil particles to relative motion.

Intrusion The emplacement of solid or molten rock in preexisting rocks.

Intrusive rock Igneous rocks deriving from the cooling and consolidation of magmas beneath the earth's surface.

Joint (or discontinuity plane) A plane (often a family of planes) which parts the rock. Other discontinuity planes of the rock mass are faults and bedding planes.

Lahars Indonesian term for mudflows defining the fluid mix of water and volcanic ejecta moving along the slopes of a volcano and either triggered by heavy rains after an eruption or by steam condensation during an eruption.

Lateral spreading The horizontal movement of sediments. The phenomenon, which can be triggered in various ways, has destabilizing effects on the foundations of structures.

Latite An igneous effusive rock made of a glassy groundmass with crystals of plagioclase and potassium feldspar.

Lithosphere The 60-100 km thick layer of the earth's crust floating over the astenosphere. The lithosphere is composed of some big plates and a number of micro-plates either colliding, stretching or separating.

Mafic Igneous rocks mainly composed of ferromagnesian minerals.

Magma Mobile molten rock with the capability of extrusion or intrusion and source material of igneous rocks after solidification.

Magnitude of an earthquake The concept, which was developed by the seismologist C.F. Richter to measure of the strength of earthquake, involves the strain energy released during such a phenomenon, based on seismographic data.

Mantle The zone of the earth between the crust and the core; the outer zone of it (upper mantle) is named astenosphere.

Metamorphism Process of alteration of solid rocks mainly due to high pressure and heat.

Meteorite Mass of matter from the outer space that struck the earth's surface.

Morpho-tectonic units Geomorphological units mainly resulting from the action of tectonics.

Mudflow General term used to express the flowing of mainly fine-graded materials and water along a streambed. It is often used as synonymous of lahar.

Ophiolite A group of igneous rocks (mafic and ultramafic) previously part of the oceanic crust.

Pelite A sedimentary rock made of fine sediments. The term is often used as synonymous of mudstone.

Phreatomagmatic eruption A volcanic explosion rich in steam and gases. It occurs when magma gets in contact with ground water or when marine water penetrates along fractures reaching the volcanic chamber.

Physiography A description of landforms.

Plagioclase A group of minerals whose composition spans from albite to anorthite.

Plate tectonics A theory of earth dynamics based on the motion of lithospheric plates floating on the astenosphere.

Pluton An igneous rock body resulting from the cooling of magma intruded into preexisting materials. The term plutonic rocks is often used as a synonymous of intrusive rocks.

Porphyrite Igneous rock with a glassy or fine-grained rockmass and coarse crystals.

Pyroclastics Volcanic material ejected during the explosive phase of a volcanic eruption.

Regolith Incoherent surface material of different origin overlying the bedrock or in situ soils.

Rejuvenation The enhancement of the erosive activity of a stream as a result of tectonic upheaval or a drop of sea level.

Reverse polarity underthrusting A case of convergent subduction. It happens in the Philippines where the South China Sea Plate is subducting eastward and the Philippine Sea Plate westward, both descending underneath the Archipelago.

Rhyodacite Effusive rock (equivalent in composition to the intrusive term granodiorite) made of quartz, plagioclase and biotite.

Rigid pavement See flexible pavement.

Sandstone A sedimentary rock mainly made of cemented medium-grained quartz grains.

Schist A medium-coarse grained and foliated metamorphic rock originated by dynamic methamorphism and mainly composed of quartz and muscovite.

Sedimentary rocks Rocks resulting from the accumulation and consolidation of mainly waterborne sediments. The family, which includes sound rocks (limestones, sandstones), cemented sediments (conglomerate, mudstones) and wind-blown volcanic ejecta (tuffs), is characterized by a layered structure.

Seismic (refraction, reflection) profiles Seismic prospecting is a geophysical method based on the analysis of elastic waves generated through the explosion of dynamite or the impact of a falling mass on the ground surface. The arrival times of elastic waves generated at a fixed point are measured at increasing distances, recorded and analyzed. By combining geologic information and seismic data a geological cross-section of the investigated profile is derived. Refraction and reflection profiles are different types of seismic investigation, depending on the distance between consecutive receivers (geophones) and the type of the array.

Shales A sedimentary laminated rock mainly originating from the consolidation of clay.

Shear Zone A tabular zone where the rock has undergone crushing during deformation and shearing.

Silt A small particle (with a diameter ranging between 0.06 and 0.004 mm) forming numerous sedimentary rocks.

Strike-slip motion The relative motion of rock blocks parallel to the fault strike.

Subduction The process by which oceanic crust descends beneath a nearby plate. Sea also Benioff Zone.

Tephra A general term for the pyroclastic materials ejected from a crater during an eruption.

Thyphoon A tropical cyclone.

Tonalite An intrusive rock made of quartz, plagioclase and horneblende.

Tornadoes A violent storm common in W Africa and U.S.

Transcurrent Fault A fault in which the two blocks shift laterally one past the other.

Transform fault A strike-slip fault typical of mid-oceanic fractures and along which the ridges are offset.

Trench A deep, steep-sided, narrow and elongated sea-floor depression usually exceeding 6 km and marking the zone between the continental margin and the abyssal plains.

Tropical cyclone Tropical cyclones, typhoons, or hurricanes are the names given to the same phenomenon in different parts of the world. They are weather systems with strong winds that circulate anti-clockwise around a low-pressure area in the northern emisphere and clock-wise in the southern emisphere. They are capable of causing massive destruction in three ways: by high winds, heavy rainfall causing inland flooding and storm-surge flooding (Source: Disaster Mitigation in Asia and the Pacific, ADB, 1991a).

Trough A flat-floored depression shallower than a trench.

Tsunami A huge marine wave produced by a volcanic eruption, earthquake or undersea avalanche.

Tuff A pyroclastic deposit made of volcanic ash and dust.

Typhoon See tropical cyclone.

Unconformity A gap in the stratigraphic sequence. Uncomformably is said of the geologic condition of layers whose contact is characterized by an unconformity.

Weathering The group of processes affecting the rocks as a result of their exposure to the atmosphere.

SUMMARY OF THE BOOK IN ITALIAN

Il libro «Geological Disasters in the Philippines» trae spunto dalla presenza dell'Autore nell'arcipelago delle Filippine durante il terremoto del 16 Luglio 1990 e la eruzione vulcanica del Pinatubo nel Giugno 1991.

L'aspetto scientifico più importante di questi fenomeni geologici è lo stretto rapporto di causa ed effetto tra terremoto ed eruzione. Per quanto concerne invece le conseguenze immediate dei due disastri e gli effetti primari e secondari, a medio e lungo termine, gli eventi geologici degli anni 1990-91 rappresentano un caso unico di successione di calamità con profonde implicazioni dovute alle interazioni tra tettonica e clima e tra questi ultimi e lo sviluppo umano.

La collisione tra lo scudo euro-asiatico e la placca del Pacifico è all'origine del terremoto del Luglio 1990 in Luzon. La tettonica delle Filippine può essere sinteticamente descritta come: a) una doppia subduzione dei fondi oceanici ad Occidente ed Oriente del Paese che si immergono in convergenza sotto l'arcipelago dando luogo a numerosi terremoti, e, b) lo scorrimento orizzontale lungo la faglia trascorrente denominata «Philippine Fault». Quest'ultima è considerata come il meccanismo attraverso il quale viene parzialmente assorbito l'accorciamento crostale prodotto dalla doppia subduzione. Una parte dell'accorciamento si traduce, infatti, nel continuo innalzamento della catena montuosa di Luzon Occidentale, denominata Cordillera Central. L'attiva erosione presente in quest'ultima, i versanti molto acclivi e l'attivo trasporto solido durante la stagione delle piogge sono in armonia con il quadro tettonico della zona.

Il terremoto che colpì l'isola di Luzon il 16 Luglio 1990 (epicentro vicino Rizal, Magnitudine 7.7 Richter ed ipocentro a 24.8 km di profondità) produsse una vistosa rottura del terreno per circa 120 km con uno scorrimento orizzontale massimo di 6.2 m ed un verticale variabile da 0.5 a 2.2 m. Rotture altrettanto estese si verificarono ad Ovest della rottura principale ma con scorrimenti minimi. Il terremoto produsse una liquefazione a livello regionale (1000 km²) nella piana denominata Central Plain ed un numero elevatissimo di frane superficiali nei rilievi della Cordillera Central e delle Caraballo Mountains. Parte del Central Luzon e la zona ad Ovest e Nord della rottura principale furono interessate dai distruttivi effetti del terremoto. Strutture di ogni tipo furono irreparabilmente danneggiate (palazzi, strade, ponti etc.) mentre le vittime furono valutate in 1666, oltre ad alcune migliaia di dispersi.

Il terremoto fu seguito da uno sciame sismico che si protrasse per alcuni mesi con intensità massima tra il Luglio ed l'Ottobre del 1990. Durata e sviluppo dello sciame, ubicazione e concentrazione degli epicentri furono attribuiti alla vasta riorganizzazione dei blocchi crostali di Luzon indotta dal terremoto del Luglio 1990.

Mentre lo sciame sismico si spostava in prevalenza verso Nord, nell'Aprile del 1991 si verificava la prima esplosione del Pinatubo. Frattanto si era anche attivato in Giappone Monte Unzen, probabilmente anche esso a seguito della rottura della Faglia della Filippine, fenomeno quest'ultimo certamente tra i più grandi di questo secolo e forse di questo millennio.

La eruzione del Pinatubo, che è ubicato nel Central Plain 110 km a NW di Manila e 90 km circa ad W dell'epicentro del terremoto del Luglio 1990, produsse ulteriori danni e morti nella zona. Con la fase parossistica della esplosione a metà Giugno 1991 il vulcano depositò circa 7 km cubici di prodotti piroclastici intorno al cono e ne disperse alcuni nell'atmosfera. Il danno derivante dalla deposizione della coltre di ceneri fu gravissimo per le cittadine ed i centri abitati della zona. La gran maggioranza delle case in Filippine sono costituite da strutture leggere i cui tetti devono solo proteggere dalla pioggia, pertanto il peso delle ceneri provocò la distruzione di numerose abitazioni. Tuttavia anche strutture più robuste ne risentirono ampiamente. La coltre di ceneri provocò la immediata distruzione dei raccolti e la paralisi delle attività agricole ed industriali portando al collasso l'economia locale.

Durante la fase critica della eruzione (12-15 Giugno 1991) iniziarono le piogge monsoniche. Queste ultime, come era logico attendersi, produssero la fluidificazione del manto di ceneri provocando vistosi «lahars», parola indonesiana con la quale vengono chiamate le colate di fango che si attivano a seguito della mobilizzazione, da parte di piogge intense, di prodotti piroclastici da poco depositati. I lahars, incanalandosi lungo i fiumi che radialmente si diramano dal cratere del Pinatubo, portarono a valle quantitativi enormi di sedimenti (oltre mezzo km cubico/anno nel periodo 1991-93), con ulteriori danni per l'agricoltura e con una vistosa modifica della rete idrografica di valle. Mentre il danno maggiore della coltre di ceneri si era attestato nel raggio di 20-30 km dal cratere, i lahars portarono ulteriore distruzione fino a 50 km.

Concludendo, i due disastri in successione ed i fenomeni ad essi legati, in parte esasperati dalla estesa deforestazione della zona, produssero danni enormi di fatto portando l'economia dell'intero stato delle Filippine a crescita zero dopo alcuni anni di ripresa.

Il libro dedica una certa attenzione al problema dei disastri naturali a livello globale nel primo capitolo. Il secondo ed il terzo capitolo sono rispettivamente dedicati ai disastri che comunemente colpiscono l'arcipelago ed alla geologia e tettonica delle Filippine. I capitoli dal quarto all'ottavo riguardano in successione il terremoto del 1990, la liquefazione nel Central Plain, le frane nelle catene Cordillera Central e Caraballo Mountains, lo sciame sismico e la eruzione del Pinatubo. Gli ultimi due capitoli sono dedicati al danno economico arrecato alla economia delle Filippine (Capitolo 9) ed agli insegnamenti appresi dai due fenomeni, nonché ad alcuni suggerimenti (Capitolo 10).

THE GEOLOGICAL TIME-SCALE

Eras	Periods	Epoches
Cenozoic	Quaternary 1.8 my - today	Holocene *
		Pleistocene
	Tertiary 65 - 1.8 my **	Pliocene 3.2 my
		Miocene 19 my
		Oligocene 14 my
		Eocene 16 my
		Paleocene 9 my
Mesozoic 225-65 my	Cretaceous 80 my	
	Jurassic 45 my	
	Triassic 35 my	
Paleozoic 570-225 my	Permian 55 my	
	Carboniferous 65 my	
	Devonian 50 my	
	Silurian 35 my	
	Ordovician 70 my	
	Cambrian 70 my	

* Holocene: 15,000 years.
** my: million years.

EARTHQUAKE SOURCE ZONES OF THE PHILIPPINES

from S.S. Su (1988), Seismic Hazard Analysis for the Philippines, Natural Hazards 1. Reprinted by permission of Kluwer Academic Publishers.

Fault Zones 1 to 8

Zone 1. Focal mechanisms for Zone 1 are predominantly of the thrust type while those of Zone 4 (which is contiguous to 1A and 1C) are generally of the strike-slip type (Lewis and Hayes, 1983, citing Cardwell et al., 1980; Seno and Kurita, 1978; Fitch, 1972; Katsumata and Sykes, 1969). Negative free-air gravity anomaly in 1A and 1C is interpreted as indicating crustal downwarping and a precursor to rupture (Lewis and Hayes, 1983). High seismicity, focal mechanism solutions, and seismic sounding profiles show Recent (Pleistocene) to Present underthrusting in 1C (Cardwell et al., 1980). However, the absence of active volcanism, and the lack of a well-defined Benioff zone and of earthquakes deeper than 200 km show that current subduction is young.

Zones 2 and 3. The two zones comprise the Philippine Trench (2B and 3B) and its Quaternary volcanic forearc (2A and 3A) (Divis, 1980). North latitude 12 degrees separates Zone 2 from Zone 3. Zone 2 has a poorly developed Benioff zone that dips westward to a depth of about 100 km (Cardwell et al., 1980). Zone 3 has a better developed Benioff zone that extends to a depth of about 200 km south of Samar, as shown by intermediate earthquakes. Considerably deeper earthquakes are found east of Mindanao and southward to Talaus Island. The relative shallowness of the dipping lithosphere in Zones 2 and 3, together with the lack of significant Quaternary volcanism in Eastern Mindanao and Samar (Lewis and Hayes, 1983), coupled with the evidence from seismic reflection profiles showing no well-developed accretionary prism in the forearc region (Hamilton, 1979; Karig, 1975; Karig and Sharman, 1975) strongly suggest that the present subduction episode may have begun only in Quaternary time, and is still propagating southward to the east of Talaud and Halmahera islands (Cardwell et al., 1980; Murphy, 1983).

Zone 4. Zone 4 is the double forearc associated with the Manila Trench which lies west of Luzon (Lewis and Hayes, 1983; Cardwell et al., 1980; Karig, 1973). 4B and 4D are the 'inner' volcanic forearc while 4A and 4C make up the 'outer' nonvolcanic forearc. Seismicity is much higher in 4C than in 4D; also relatively higher in 4 A than in 4B. 4B and 4D are marked by presently active volcanoes and Quaternary cones (Cardwell et al., 1980). Focal mechanism solutions of earthquakes in the North Luzon Ridge (4B) are predominantly of the strike-slip type. Some of these are left-lateral and some right-lateral.

Zone 5. Zone 5 is the Manila Trench. It has a Benioff zone that dips 40 degrees eastward. The dip angle increases to almost 90 degrees as one proceeds southward to Manila Bay. The Benioff zone extends to a depth of about 200 km (Hayes and Lewis, 1984). South of latitude 13 degrees, the Manila Trench changes its trend from north-south to northwest-southeast, and curves towards the Mindoro Strait. This is believed to be due to the collision of the subducting lithosphere with the North-Palawan (Calamian) micro-continental block (Lewis and Hayes, 1984). Seismicity is lower in the trench itself (5A) than in the forearc (5B). Present day volcanic activity in the forearc is an indication of convergence between the South China Sea Plate and Luzon (Hamburger et al., 1983). Convergence rate is estimated at about 10 to 20 mm per year (Hayes and Lewis, 1984).

Zone 6. Zone 6 comprises the Negros Trench and the Sulu Trench (6A) and their volcanic forearc (6B). Seismicity is relatively low in this subduction zone. There are relatively few shallow earthquakes of the thrust type in the Negros Trench. However, seismic refraction profiles show sediments being underthrusted to the east along the Negros Trench (Cardwell et al., 1980; Hamilton, 1979). There is no clearly defined Benioff zone; but there are intermediate earthquakes suggesting that the lithosphere is being subducted eastward (Cardwell et al., 1980). The Sulu Ridge (6B) parallels the Palawan Ridge which has no seismicity at all. The Sulu Ridge is associated with a subduction that occurred from the Late Cenozoic to Pleistocene (Cardwell et al., 1980). The Negros Trench and the Sulu Trench are considered to be one tectonic unit (Divis, 1980). Kanlaon is an active volcano on Negros island.

Zones 7 and 8. Zone 7 is a shallower structure, while Zone 8 is the deeper structure and is the northern extension of the Molucca Sea Plate. This plate had buckled and dips both westward and eastward. Its surface expression is the Sangihe Ridge. Zone 8 is the westward dipping portion of the Molucca Sea Plate. It is characterized by intermediate earthquakes (8B) and deep earthquakes (8A), as far as 680 km (Cardwell et al., 1980).

Fault Zones 9 to 18

Following the division of Philippine faults into transcurrent, normal and thrust faults according to the Philippine Bureau of Mines (1981), three transcurrent, two normal, and thrust faults are considered very probably seismically active. In addition, two transform faults are added. Thus, a total of ten faults are selected as source zones. In the contest of the methodology commonly in use, there are two possible ways of viewing faults. One is to regard them as line sources; the other is to treat them as finite-width sources. In regarding them as line sources, as for example in McGuire's 'Frisk' program (1978), additional fault parameters such as rupture length have to be considered. In treating them as finite-width sources, their parameters are no different from those of zones 1 to 8 earlier described. A test was made to compare the results of treating a fault as a line source and as a finite-width source. The results agree very closely. Thus, as a measure of convenience, faults are considered finite-width sources and included in the same computation used for Zones 1 to 8.

Zone 9. Zone 9 is the transcurrent Philippine Fault that extends over 1200 km from Lingayen Gulf in Luzon to Davao Gulf south of Mindanao (Allen, 1962). Krause (1966) suggests that it extends south of Mindanao along a submarine scarp into the Talaud Islands. Movement is left-lateral strike-slip (Ranneft et al., 1960; Allen, 1962; Rutland, 1968). Morante and Allen (1974) studied the geomorphic effects of the 1973 Regay Gulf earthquake and found a left-lateral displacement of 3.2 m near the Tayabas Isthmus of Southern Luzon. This was later confirmed by an earthquake focal mechanism solution of Lewis and Hayes (1983). Other earthquakes (1937, 1973, 1975) studied by Acharya and Aggarwal (1980) have focal mechanisms that can be correlated with the left-lateral movement along the fault. Geomorphic features in the other parts of Luzon, Masbate and Leyte likewise indicate left-lateral movements (Allen, 1962). Eight other large earthquakes, namely, of 1893, 1901, 1911, 1924, 1937, 1941, 1947, 1948, had their epicenters along or very near the fault zone (Rowlett and Kelleher, 1976). Thus seismicity gives evidence that the Philippine Fault is presently active, at least in some parts. Additional evidence from geology, such as the sharpness of fault scarps, disrupted soil horizons and stream offsets confirmed that the fault has been active since Quaternary time (Lewis and Hayes, 1983). Certain surface deformations near and along the fault have been associated with historical earthquakes such as those of 1989, 1879, and 1983 (Allen, 1962). In Central Luzon, the fault divides the mountainous Cordillera Central in the north from the lowlands of the Central Valley Basin in the south. Past episodes of intense activity along the fault have been placed in Late-Miocene and Post-pliocene (Rutland, 1968).

Zone 10. Zone 10 is another transcurrent fault, the Tablas Lineament that is treaceable for about 350 km from Western Panay northward through Tablas Island to the Tayabas Isthmus (Phil. Bureau of Mines, 1981). Allen (1962) pointed out it might be conjugate to the Philippine Fault and thus, right-lateral.

Zone 11. Zone 11 is the third transcurrent fault, called Mindanao Fault. It is traceable for about 400 km, from the Davao Gulf northwestward to the Sindangan Valley of the Northern Zamboanga (Phil. Bureau of Mines, 1981; Gervasio, 1964). On its northwest end there is a possibility it might be a high-angle thrust; but it is believed to be more probably a transcurrent fault. There is a fourth transcurrent fault, the Ulugan Fault in the Island of Palawan, but it is considered inactive because its seismicity is practically nil.

Zones 12 and 13. Zone 12 is a normal fault on the northeast side of Mindanao. Zone 13 is the Cotabato Normal Fault on the southwest flank of Mindanao. Another normal fault, the Marikina Fault, located east of Manila, is not included because it is considered inactive.

Zones 14 to 16. Zone 14 is the Zambales Thrust Fault on the western part of Luzon, that is north of Manila Bay. Zone 15 is a thrust fault on the southwest side of Mindoro. Zone 16 is a thrust fault on the western flank of Panay Island. The series of thrust faults along Zamboanga peninsula are not included here because they are considered to be inactive.

Zones 17 and 18. Zone 17 is a transform fault at about north latitude 15.5 degrees linking the East Luzon Trench (Zone 1) with the Philippine Trench (Zones 2 and 3) (Lewis and Hayes, 1983; Hamburger et al., 1983). Confirmatory evidence consists of concentrated seismicity, sharp bathymetric low, and focal mechanism solutions of two earthquakes (seno and Kurita, 1978; Cardwell et al., 1980). Zone 18 is a transform fault running along the Verde Island passage north of Mindoro. Supposedly it links the Manila Trench and the Negros Trench. This interpretation finds confirmation in the intense or concentrated seismicity north of Mindoro (Wolfe and Self, 1983). Its sense of movement is believed to be left-lateral (Lewis and hayes, 1984).

THE ROSSI-FOREL SCALE OF EARTHQUAKE INTENSITIES
(SEEASE, 1985)

I.   Hardly perceptible shock - felt only by an experienced observer under favourable conditions.

II. Extremely feeble shock - felt by a small number of persons at rest.

III. Very feeble shock - felt by several persons at rest. Duration and direction may be perceptible. Sometimes dizziness or nausea expereinced.

IV. Feeble shock - felt generally indoors, outdoors by a few. Hanging objects swing slightly. Creaking of frames of houses.

V.   Shock of moderate intensity - felt generally by everyone. Hanging objects swing freely. Overturning of all tall vases and unstable objects.

VI. Fairly strong shock - general awakening of those asleep. Some frightened persons leave their houses. Stopping of pendulum clocks. Oscillation of hanging lamps. Slight damage to very old or poorly-built structures.

VII. Strong shock - overturning of movable objects. General alarm, all run outdoors. Damage slight in well-built houses, considerable in old or poorly-built structures, old walls, etc. Some landslides from hills and steep banks. Cracks in road surfaces.

VIII. Very strong shock - people panicky. Trees shaken strongly. Changes in the flow of springs and wells. Sand and mud ejected from fissures in soft ground. Small landslides.

IX. Extremely strong shock - panic general. Partial or total destruction of some buildings. Fissures in ground. Landslides and rock falls.

www.ingramcontent.com/pod-product-compliance
Lightning Source LLC
Chambersburg PA
CBHW051017180526
45172CB00002B/384